W9-CPE-196

HANDBOOK OF RIGGING

OTHER McGRAW-HILL HANDBOOKS OF INTEREST

ABBOTT AND STETKA · National Electrical Code Handbook
AMERICAN SOCIETY OF MECHANICAL ENGINEERS · ASME Handbooks:
    Engineering Tables   Metals Engineering—Processes
    Metals Engineering—Design   Metals Properties
ARCHITECTURAL RECORD · Time-Saver Standards
BELL · Petroleum Transportation Handbook
BRADY · Materials Handbook
CHOW · Handbook of Applied Hydrology
CONOVER · Grounds Maintenance Handbook
CROCKER · Piping Handbook
CROFT AND CARR · American Electricians' Handbook
DAVIS · Handbook of Applied Hydraulics
FACTORY MUTUAL ENGINEERING DIVISION · Handbook of Industrial Loss
    Prevention
FLÜGGE · Handbook of Engineering Mechanics
FRICK · Petroleum Production Handbook
GUTHRIE · Petroleum Products Handbook
HARRIS · Handbook of Noise Control
HARRIS AND CREDE · Shock and Vibration Handbook
HEYEL · The Foreman's Handbook
KATZ · Handbook of Natural Gas Engineering
KING AND BRATER · Handbook of Hydraulics
KNOWLTON · Standard Handbook for Electrical Engineers
KORN AND KORN · Mathematical Handbook for Scientists and Engineers
LA LONDE AND JANES · Concrete Engineering Handbook
LASSER · Business Management Handbook
LAUGHNER AND HARGAN · Handbook of Fastening and Joining of Metal Parts
MAGILL, HOLDEN, AND ACKLEY · Air Pollution Handbook
MANAS · National Plumbing Code Handbook
MANTELL · Engineering Materials Handbook
MARKS AND BAUMEISTER · Mechanical Engineers' Handbook
MAYNARD · Top Management Handbook
MOODY · Petroleum Exploration Handbook
MORROW · Maintenance Engineering Handbook
MULLIGAN · Handbook of Brick Masonry Construction
PERRY · Engineering Manual
ROSSNAGEL · Handbook of Rigging
STANIAR · Plant Engineering Handbook
STREETER · Handbook of Fluid Dynamics
STUBBS · Handbook of Heavy Construction
TIMBER ENGINEERING Co. · Timber Design and Construction Handbook
URQUHART · Civil Engineering Handbook
VON BERNEWITZ · Handbook for Prospectors
WOODS · Highway Engineering Handbook
YODER, HENEMAN, TURNBULL, AND STONE · Handbook of Personnel Management
    and Labor Relations

Handbook of Rigging

For Construction and Industrial Operations

W. E. ROSSNAGEL, P.E.

Consulting Safety and Fire Protection Engineer
Formerly Safety Engineer,
Consolidated Edison Company of New York, Inc.

THIRD EDITION

McGRAW-HILL BOOK COMPANY

New York   Toronto   London

HANDBOOK OF RIGGING

16 – MAMM – 7 5

53940

PREFACE

The *Handbook of Rigging* is intended as a ready reference and guide for expert riggers who are engaged full time in their work, for erectors of buildings and structures, for maintenance mechanics in industrial plants and electric generating stations who have less frequent rigging jobs to perform, for operators of all types of hoists, derricks, and cranes, for painters and masons who have scaffolds to erect, for erectors of signs, and for steeplejacks, whose lives depend upon the ropes and blocks which they use. The construction engineers, superintendents, and foremen who have jurisdiction over these trades should also find this book useful.

For the benefit of the practical worker an attempt has been made to make reading easy, for this book is written in plain, not too technical language. Some workmen may be interested in mathematics, so some simple formulas have been included for calculating the safe loads on hoisting tackle, slings, steel and timber beams, and the like. To make these formulas more readily understood, a number of practical problems have been worked out. Thus the strength of equipment can be estimated much more accurately than by rule-of-thumb or hit-or-miss methods.

In the Appendix are mathematical and other tables which may be handy for the design engineer or construction engineer. Many material inspectors in the purchasing departments of large industrial corporations have found the text very useful when inspecting such materials as wood ladders, scaffold planks, manila rope, and wire rope.

Special types of scaffolds and numerous newly developed kinds of equipment are shown in the illustrations. If such devices are not on the market at this time, they may be constructed in plant shops, using the detail drawings provided.

The Third Edition of the *Handbook of Rigging* has been revised, enlarged, and brought up to date. In addition to the lifting or

hoisting aspects of rigging, mention is made of moving loads horizontally—transportation.   While the rigger does not actually
drive the truck or semitrailer, he has the responsibility for loading it in such a manner that the cargo will ride safely, and he
has to remove the load at its destination.   Inasmuch as a sur-
prisingly large number of copies of earlier editions have been sold
or taken into underdeveloped countries where roads may be
scarce, mention is made of the "overland train" which can
transport heavy loads at fair speed over desert sands and arctic
glaciers, or through jungle swamps.

Since publication of the First Edition, with its emphasis upon
smaller buildings, so many requests have been received from
readers for broader coverage that an effort has been made to
include problems associated with the erection or demolition of
taller buildings, including the lifting of heavier loads.   Consequently, more space has been devoted to erection derricks and
to the determination of safe working stresses involved in their
operation.   Technical data on the strength and safety of tubular
steel scaffolding have also been added, together with information
about some newer types of mechanical splices for wire rope and
instructions for making hand splices.

Techniques and methods have been suggested to accomplish
the rigger's tasks with the greatest safety for all the men at work
on a project, as well as for passers-by and the public in general.
Because the electrical hazard from accidental contact of the
boom or cable of automotive cranes with overhead power lines
has resulted in an alarming increase in fatalities, less reliance has
been placed on the use of inadequate grounding wires and more
on the necessity for avoiding such contact.   As has often been
said, "safety is no accident"—and a safely performed job can be
achieved by taking the necessary precautions and by proper
training and supervision.   The cost of neglect is measured not
only in terms of occasional injury to individuals, but in those less
spectacular ways—lost time, damage to materials and work
in place, and high job costs generally—that destroy working
economy and efficiency.

Most of the chapters in the *Handbook of Rigging* are based
upon data accumulated by the author during his years as a
safety engineer specializing in construction operations and incorporated in his many articles published in such technical magazines
as *Construction Methods, Engineering News-Record, Power, Factory*

Management and Maintenance, Southern Power and Industry, Power Plant Engineering, Ingenieria e Industria, Supervision, Safety, Safety Engineering, and *National Safety News.*

The author's sincere thanks are extended to the following gentlemen: Marshall Olds of the Jones and Laughlin Steel Corporation; Capt. A. R. Hatch of the Columbian Rope Company; G. L. Griffith of the Cable Division, and A. J. Gugliotta of the Chain Division, American Chain and Cable Company, Inc.; W. W. Weber of the Forest Products Laboratory of the U.S. Department of Agriculture; H. Richardson of John A. Roebling's Sons Division, the Colorado Fuel and Iron Corporation; H. S. Gehr of The Patent Scaffolding Company, Inc.; L. H. Davidson of Chesebro-Whitman Company, Inc.; W. E. Clapp of the Yale and Towne Manufacturing Company; W. J. Goetz, A. Nacht, T. J. Shaughnessy, and Dr. N. E. Eckelberry of the Consolidated Edison Company of New York, Inc.; and last, but certainly not least, the late Prof. Lionel S. Marks of Harvard University, who so kindly granted permission to use certain tables from his *Mechanical Engineers' Handbook.*

Among the companies and organzations that furnished drawings, photographs, and photoengravings to help make this handbook more interesting and instructive are Jones and Laughlin Steel Corporation; National Park Service; Bucyrus-Erie Company; Parke-Thompson Associates; Climbing Tower Derrick Company; American Chain and Cable Company, Inc.; American Steel and Wire Division, United States Steel Corporation; John A. Roebling's Sons Division, the Colorado Fuel and Iron Corporation; Columbian Rope Company; Encyclopaedia Britannica, Inc.; Yale and Towne Manufacturing Company; Mac-Whyte Company; Safway Steel Products, Inc.; Chesebro-Whitman Company, Inc.; The Patent Scaffolding Company, Inc.; American Standards Association; National Safety Council; Forest Products Laboratory; Harnischfeger Corporation; Rose Manufacturing Company; and R. G. Le Tourneau, Inc.

W. E. Rossnagel

CONTENTS

CHAPTER I

FUNDAMENTALS

The art of rigging may be traced back to prehistoric times. Levers were used then, as now, to pry stones, roll logs, and move objects that were too heavy to be moved by hand. The inclined plane, or a natural ramp, has been used since time immemorial to help move heavy objects up to higher elevations.

All present-day riggers should be interested in the first major rigging job of which there is not only a record, but also indisputable evidence. It was the construction, about 2700 B.C., of the three pyramids at Gizeh, near Cairo, Egypt. The largest of these was built to contain the remains of the Pharo Cheops. It is believed that the preliminary work prior to actual construction required about 10 years. Actual construction is said to have required about 20 years.

As it stands today the large pyramid is 746 ft square at the base and 451 ft high. It contains about 2,300,000 stones weighing from 2 to 30 tons each, and totaling about 5,750,000 tons, nearly 20 times the weight of the masonry in the 102-story Empire State Building in New York City. As built, the stone structure was encased in a sheath of fine grain limestone, but at some unknown time during the past 4,600 years this sheath was removed.

It is said that these huge blocks of stone were transported from the quarries to the bank of the Nile River. Then, during the annual 3-month period when the river was at its flood stage, and therefore very wide, they were ferried across the Nile. And again they were dragged to the construction site. History records that a sand ramp was built up one side as the pyramid rose in height, and eventually reaching to its apex. This ramp is said to have required nearly a million tons of sand to be transported from the desert and piled up at the site. In addition, another million tons of sand were used to backfill the interior of the pyramid. And when the job was completed the ramp had to be removed and distributed over the desert area.

1

The construction crews had no automobile trucks, no railroads, earthmoving machinery, automotive cranes, derricks, rope falls, chain hoists, jacks, or other mechanical equipment. They are believed to have used levers, rollers, crude ropes, sledges, plumb lines, and string sighting to do their work. Nevertheless, as incredible as it may seem, the pyramid remains today as indisputable proof of the ingenuity of the Egyptians. We are told that these huge stones were hauled up the ramp on rollers by the brute strength of 100,000 slaves in teams of 50 men each, driven by the whips of the slave masters.

The average height of lift of the stones may be assumed to be about 100 ft. To fully appreciate the enormity of this undertaking let us consider the power developed by men.

Turning a crank, such as on a winch, a man may exert 15 to 18 lb force continuously. Intermittently he may exert 25 to 40 lb. Pulling downward on a rope or on the hand chain of a chain hoist he may be able to exert 40-lb pull for a long time, but for a short period his pull may approach his weight. In lifting a few inches off the ground (assuming that he does it in a proper manner) he may lift up to 300 lb. Pushing or pulling an object, such as a vehicle, he may with good footing exert a force of 110 to 130 lb.

But let us look into the amount of work that he may be expected to accomplish in an 8-hour day, and then compare his effort with the cost of doing the job electrically. If electric power costs, let us say, 3 cents per kilowatt-hour, and assuming that the electrically driven apparatus is 50 per cent efficient, then the cost equivalent of the man's labor may be somewhat near the figures given below: Keep in mind that for continuous work a man may be expected to deliver about 0.10 hp, while for a very short time he may exert from 0.4 to 0.5 hp.

A strong man can lift 86 tons (say bags of cement) from the ground to a height of 4 ft in an 8-hour work day, averaging 0.045 hp. The cost of doing this electrically would be about 1.6 cents per day.

A man can carry 22.3 tons up a ramp or stair to a height of 12 ft in an 8-hour day, averaging 0.034 hp. Electrical cost 1.2 cents.

Pushing a wheelbarrow he can move 40.7 tons up a 3-ft ramp in 8 hours, and averages 0.015 hp. Power cost 0.52 cents.

Shoveling loose earth he can raise 20 tons to shoulder height in 8 hours, exerting 0.013 hp. Equivalent power cost 0.45 cents.

Thus, it will be noted that man is a very inefficient machine, and it is easy to understand why 100,000 men were required to transport the material for the Great Pyramid.

The art of rigging has developed to the degree that today we think nothing of building 600-ton traveling cranes for power plants and hammer-head cranes of 300 tons or greater capacity for Navy yards. Trusses and girders weighing up to 50 or 100 tons or more are handled by derricks when erecting modern office buildings or bridges. During the Second World War hundreds of steam locomotives weighing over 100 tons each were hoisted onto the decks or into the holds of steamships for transportation to Allied countries.

The author did not intend that this book should deal with rigging operations of this magnitude, but rather with everyday maintenance operations in industrial plants, in factories, in power plants, in the transportation and handling of heavy machinery, and in the erection and demolition of smaller size structures.

However, numerous requests for additional information on this phase of rigging have been received by the author, especially from Europe and the Near East where they apparently are not familiar with American building erection practices. Consequently, more reference is made in this edition of the *Handbook* to the erection of tall buildings.

Rigging includes a multitude of subdivisions, which are included in subsequent chapters. Let us imagine that a heavy piece of machinery is to be hoisted up the outside of a building and in through a window opening. Dependent upon the weight of the load, either a fiber rope or a wire rope tackle will be needed. This may be suspended from a timber gin pole or from a wood or steel outrigger temporarily projected out from a window above or from the roof. To attach the load to the hook on the lower pulley block, a sling is required. Then to support the load as it is swung in through the opening in the wall, a chain hoist may be suspended from a beam of the floor above. When the load is landed, it will probably be moved across the floor on rollers and propelled by pinch bars. Thus, in order to do this relatively simple rigging job, rope, hooks, slings, timbers, a chain hoist, rollers, etc., may be called into use.

Rigging also includes erecting temporary scaffolds for painting, repairing, or demolishing structures as well as for supporting heavy loads. In the rigging trade there is no substitute for years of practical experience. But when working by rule of thumb alone, many times ropes and other load-bearing parts are stressed to a point dangerously close to the breaking point without the rigger realizing it.   The author has no intention of attempting to make professional engineers out of riggers, but owing to the fact

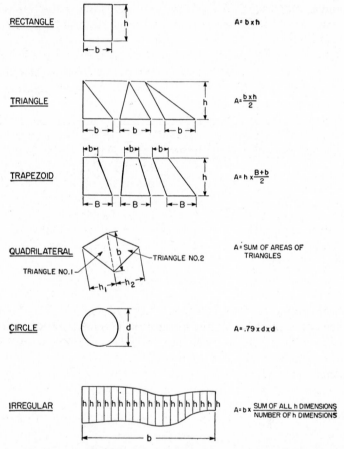

RECTANGLE $A = b \times h$

TRIANGLE $A = \dfrac{b \times h}{2}$

TRAPEZOID $A = h \times \dfrac{B+b}{2}$

QUADRILATERAL   TRIANGLE NO.2 $A = $ SUM OF AREAS OF TRIANGLES
TRIANGLE NO.I

CIRCLE $A = .79 \times d \times d$

IRREGULAR $A = b \times \dfrac{\text{SUM OF ALL h DIMENSIONS}}{\text{NUMBER OF h DIMENSIONS}}$

FIG. 1.   Formulas for calculating the areas of plane figures.

that most riggers today have had better educational advantages than in years gone by, they should be able to make simple calculations to check their loads and the strength of their equipment. The element of chance should be thus reduced to a minimum.

The first and the most important step in any rigging operation is to determine the actual weight of the load to be supported or hoisted. If this information cannot be obtained from shipping papers, from catalogue data, or from other dependable sources, it may be necessary to roughly calculate the weight. If the object is of very complex construction, it may be necessary by observing it to estimate the size of a solid chunk of metal it could be melted into and multiplying by the weight per cubic foot of that metal. Such estimate will be very inaccurate but probably will be somewhat better than a pure guess.

If the object is of more simple shape, imagine it cut up into a number of regular geometric figures, the cubic contents (and the weight) of which can be calculated with a fair degree of accuracy. The sum of these figures will give the total weight.

Figure 1 gives the formulas for calculating the area of a few plane figures frequently encountered. For example, to find the area of a flat plate of irregular shape (Fig. 2a), draw chalk lines on

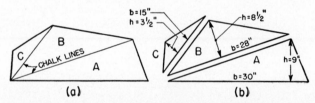

FIG. 2. Calculating the area of a complex plane figure.

the plate from any corners desired so as to subdivide it into a number of triangles. Take the necessary measurements and apply the formula for triangles given in Fig. 1. The sum of the triangles equals the area of the plate.

In this example Fig. 2b shows the shape cut into three triangles *A*, *B*, and *C*. Triangle *A* has a base of 30 in. and a height of 9 in., and its area $= \dfrac{b \times h}{2} = \dfrac{30 \times 9}{2} = 135$ sq in. Triangle *B* thus has an area of $\dfrac{28 \times 8\frac{1}{2}}{2} = 119$ sq in., and triangle *C* has an area of $\dfrac{15 \times 3\frac{1}{2}}{2} = 26\frac{1}{4}$ sq in. Thus, the area of the plate is 135 plus 119 plus $26\frac{1}{4}$ sq in., or a total of $280\frac{1}{4}$ sq in. This area multiplied by the thickness of the plate, say $\frac{3}{4}$ in., gives 210 cu in. volume. From Table I we learn that steel has a weight of 0.28 lb per cu in. The weight of the plate is found to be 210 cu in. times 0.28 lb, or 59 lb.

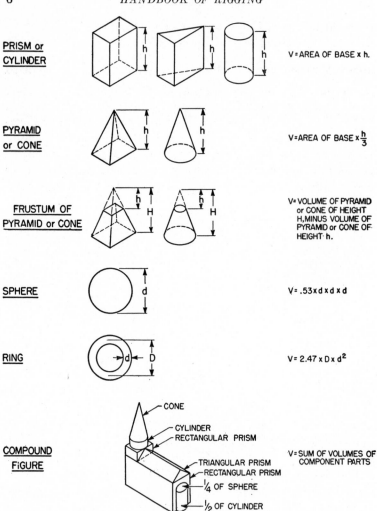

PRISM or CYLINDER   V = AREA OF BASE x h.

PYRAMID or CONE   V = AREA OF BASE x $\frac{h}{3}$

FRUSTUM OF PYRAMID or CONE   V = VOLUME OF PYRAMID or CONE OF HEIGHT H, MINUS VOLUME OF PYRAMID or CONE OF HEIGHT h.

SPHERE   V = .53 x d x d x d

RING   V = 2.47 x D x d²

COMPOUND FIGURE   V = SUM OF VOLUMES OF COMPONENT PARTS

CONE
CYLINDER
RECTANGULAR PRISM
TRIANGULAR PRISM
RECTANGULAR PRISM
¼ OF SPHERE
½ OF CYLINDER

FIG. 3.   Formulas for calculating the volumes of solid figures.

Figure 3 illustrates the method of calculating the volume of a few elementary solid figures.   Like the plane figures, it may be necessary to subdivide the object into smaller geometric shapes and find the volumes of the individual parts.   As an example, let us estimate the weight of a concrete mass shown in Fig. 4.   First, we divide it into a rectangular block A and a frustum of a pyramid B.   The volume of part A is $4 \times 5 \times 3\frac{1}{2}$ ft $= 70$ cu ft.   To get the volume of part B, it is necessary to figure the volume of a pyramid having a height of 6 ft and subtract the volume of a pyramid of height of 3 ft. The volume of the large pyramid is the area of the base times one-third the height, or $\dfrac{4 \times 5 \times 6}{3} = 40$ cu ft. From this must be subtracted the volume of the small pyramid $\dfrac{2 \times 2\frac{1}{2} \times 3}{3} = 5$ cu ft, thus giving the frustum a volume of 40 cu ft $-$ 5 cu ft = 35 cu ft. The total volume of the concrete block is 70 + 35 cu ft = 105 cu ft. Concrete weighs about 144 lb per cu ft (from Table I) or 144 lb $\times$ 105 cu ft = 15,120 lb, or about $7\frac{1}{2}$ tons. This could have been calculated on a cubic inch

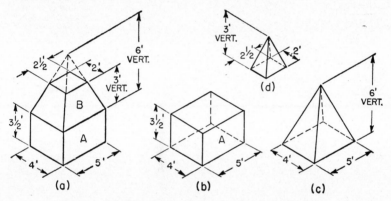

Fig. 4. Calculating the volume of a complex solid figure.

basis rather than cubic foot, and the result would have been the same.

The rigger frequently has occasion to determine the approximate location of the center of gravity of an irregular-shaped load he intends to lift. By "center of gravity" is meant the location where the center of the entire weight of the object is theoretically concentrated. This point, when the object is freely suspended from a hook, will always be directly below the hook.

Of course the center of gravity of a solid figure has to be located in three planes or directions, namely lengthwise, crosswise, and vertically. Finding the exact location requires mathematical calculations, but for the average rigging job it can be estimated closely enough for our purpose. Figure 5 shows some familiar and some irregular-shaped plane figures representing, say, the length or cross section of the object and showing the approximate locations of their centers of gravity. Most are within the object, but a few are located outside.

In Chapter XVII is shown the need for ascertaining the location of the center of gravity of a load to be hoisted.

A simple but exact method of locating the center of gravity of a plane figure of irregular shape is to cut it out, at any convenient scale, from a piece of cardboard (Fig. 6a). Punch pinholes near two adjacent corners of the cardboard and suspend it freely from a pin stuck into a wall or other vertical surface. Also

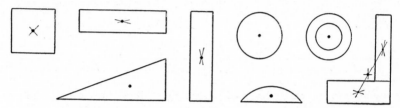

FIG. 5.   Approximating the location of the center of gravity of plane figures.

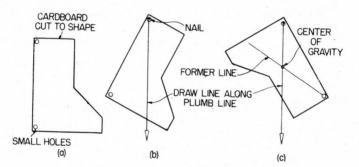

FIG. 6.   Locating the center of gravity of an irregular plane figure.

suspend from the pin a small weight attached to a string, and draw a line on the card along the string (Fig. 6b). Repeat, using the other hole. Where the two pencil lines cross (Fig. 6c) is the center of gravity.

A few words about elementary physics may also be helpful to the modern rigger. The elemental machines from which all machinery is constructed are the inclined plane, the lever, the wheel and axle (gear or pulley and shaft), and the block and fall. Riggers frequently make use of the inclined plane when hauling a load on rollers up a ramp or up the skids onto a truck. To estimate roughly the pull required to haul the load, draw a diagram

TABLE I. WEIGHTS OF SOME COMMON MATERIALS

Material	Lb per cu in.	Lb per cu ft
Wood (spruce)............................	0.016	27
Wood (longleaf pine)......................	0.025	44
Coal (bituminous).........................	0.029	50
Water....................................	0.036	$62\frac{1}{2}$
Earth....................................	0.058	100
Sand.....................................	0.070	120
Concrete.................................	0.083	144
Cast iron................................	0.24	442
Steel....................................	0.28	487
Brass....................................	0.31	534
Lead.....................................	0.44	710

ABC (Fig. 7) to scale to represent the incline of, say, 4 ft in 20 ft. A load of 15,000 lb is to be pulled up on rollers. Draw to any suitable scale a vertical line *DE* to represent the weight of the load. For instance, assume a scale of 1 in. = 10,000 lb; then a line $1\frac{1}{2}$ in. long represents $1\frac{1}{2} \times 10,000$ lb, or 15,000 lb. From *D* draw a line *DF* at right angle to the slope of the incline *AB*.

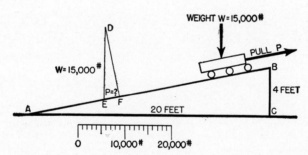

FIG. 7. Calculating the force required to move an object up an inclined plane or ramp.

Using the same scale of 1 in. = 10,000 lb, measure the distance *EF*, which will be the theoretical pull required. This scales about 0.3 in. or 3,000 lb, which is the pull if frictionless rollers are used. To this must be added the resistance due to friction.

The screw jack is an adaptation of the inclined plane, and more will be said about it in Chapter XIX.

The crowbar is a typical lever. As commonly used, there are two arrangements of the fulcrum (see Fig. 8). In Fig. 8*a* the

upward pull P on the handle lifts the weight W, which owing to
gravity is acting as a downward force.   The toe of the crowbar
pivots about the fulcrum F, and in effect, the floor or ground
exerts an upward force to resist the downward pressure.   Let us
assume a weight of 1,000 lb acting at a distance of 3 in. from the
fulcrum F and a man lifting up on the lever at a distance of 30 in.
from the fulcrum.   The force P times the force distance (30 in.)
always equals the weight (W = 1,000 lb) times the weight distance (3 in.).   Thus $P \times 30$ in. = 1,000 lb $\times$ 3 in., or P = 100 lb
force required.   In other words, if the force distance is ten times
the weight distance, then the force is one-tenth of the weight,
and this rule holds true for other ratios of distances.

If the crowbar is used as indicated in Fig. 8b, the fulcrum is
between the force and the load, and the former is then a push P in

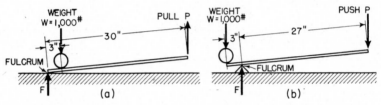

FIG. 8.   Calculating the force required to lift or move loads by means of levers.

a downward direction.   If the same 30-in. lever is used, we find
that the force P times the force distance (27 in.) equals the weight
(W = 1,000 lb) times the weight distance (3 in.).   Thus

$$P \times 27 \text{ in.} = 1{,}000 \text{ lb} \times 3 \text{ in.}$$

or P = 111 lb force required.   It will be observed that the crowbar is most efficient, or less force is required, when used as shown
in Fig. 8a.   Sometimes it may be necessary to limit the pressure
on the fulcrum F.   In Fig. 8b this pressure is always the sum of
the weight W and the force P.

The wheel-and-axle is in reality only an adaptation of the lever,
but whereas the movement of the lever is limited, the motion of
the wheel may continue indefinitely (see Fig. 9).   In this example
force P is exerted on the radius (R = 12 in.) of the large pulley
while the weight (W = 1,000 lb) is supported on radius (r = 3 in.)
of the small pulley.   Thus $P \times 12$ in. = 1,000 lb $\times 3$, or
P = 250 lb.   In making use of the wheel and axle (which in
practice is demonstrated by belts and pulleys and by gears on the

same shaft), the belt pulls or tensions are inversely proportional to the pulley diameters, or in other words, if one pulley is four times as large as the other, the belt pull on the large pulley is one-fourth of the belt pull on the small pulley.

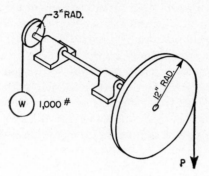

Fig. 9.   Calculating the force required to lift a load by means of the wheels and shaft.

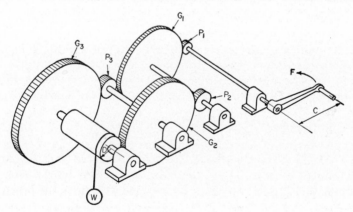

Fig. 10.   Calculating the force required to lift a load by means of a train of gears.

If there is a train of gears, such as on a hand winch (Fig. 10), consisting of several gear reductions, the theoretical pull which can be exerted on the hoist rope may be readily estimated.   Of course a deduction must be made for friction.

$$W = \frac{F \times C \times G_1 \times G_2 \times G_3 \text{ etc.}}{D \times P_1 \times P_2 \times P_3 \text{ etc.}}$$

where W = pull on hoist rope, in lb

$\quad F$ = force on hand crank, about 17 lb at $2\frac{1}{2}$ ft per second

$\quad C$ = radius of hand crank, in in.

$\quad G_1$ = number of teeth in gear G_1

$\quad G_2$ = number of teeth in gear G_2

$\quad G_3$ = number of teeth in gear G_3 (if such exists)

$\quad D$ = radius of drum, to center of rope, in in.

$\quad P_1$ = number of teeth in pinion P_1

$\quad P_2$ = number of teeth in pinion P_2

$\quad P_3$ = number of teeth in pinion P_3 (if such exists)

For a discussion of the forces concerned in a block and fall arrangement, refer to Chapter II on Fiber Rope. In each of these reevings, as well as in the winch, friction consumes a portion of the energy imparted to the mechanism; consequently the actual force P to be applied must be somewhat greater than calculated. The power wasted in overcoming friction may vary from 5 to nearly 100 per cent, depending upon the condition of the shafts, bearings, gears, pulleys, rollers, etc., and upon their lubrication.

A general knowledge of these mechanical movements will help the rigger better to understand the operation of the tools and machines that he uses in his work.

Attention of the rigger should be called to the additional stress which is placed on a rope, chain, beam, scaffold, ladder, or other load-bearing member as a result of a suddenly applied load, such as a jerk or an impact. A beam, for instance, may safely support at its center a concentrated load of say 1,000 lb. If a wheel carrying a 1,000-lb load is rolled at high speed over this beam it will produce a stress in the beam twice as great as that produced by the static (stationary) load.

If the 1,000-lb load, which we will assume for the present is incompressible, is dropped onto the beam from a height of two feet the energy developed at the instant of impact will be 1,000 lb, times 2 ft, or 2,000 ft lb. This energy must be absorbed by the beam. If we imagine that the beam could be deflected 2 ft, the impact would be $\dfrac{2{,}000 \text{ ft lb}}{2 \text{ ft}} = 1{,}000$ lb, in addition to the static load of 1,000 lb which would be applied if the load was at rest on the beam. If the beam should be deflected 1 in., then the impact will be $\dfrac{2{,}000 \text{ ft lb}}{\frac{1}{12} \text{ ft}} = 24{,}000$ lb, plus 1,000 lb, or 25,000 lb. If deflected only $\frac{1}{4}$ in. the impact will be $\dfrac{2{,}000 \text{ ft lb}}{\frac{1}{48} \text{ ft}} = 96{,}000$ lb, plus 1,000 lb, or 97,000 lb. Undoubtedly the beam will fail. Actually the load will compress slightly, and/or the sling will stretch, and consequently the stopping distance of the center of gravity of the load will be somewhat greater.

If the movement is horizontal, or nearly so, such as a heavy slab on a pair of rollers which gets out of control on a slight downgrade, and assuming that it strikes a fixed object such as a building column, the same rule will apply, except that the static weight is not added. The impact can be estimated if the vertical distance "dropped" (probably only an inch or so) is known and the deflection of the column assumed. If, on the other hand, the speed should be known, then it is possible to calculate the theoretical vertical drop by using the following formula:

$$H = \frac{V^2}{2g} \quad \text{or} \quad H = \frac{V^2}{64.32}$$

where H is the theoretical drop in feet, and V is the velocity in ft per sec.

For example, should a 3,600-lb automobile traveling 60 miles per hour (88 ft per sec) run head-on into a massive concrete bridge abutment backed up by an earth embankment, the movement of the abutment will be nil. However, let us assume that when the car comes to rest the rear axle is 3 ft nearer to the wall than when the front bumper first touched it. In other words, the center of gravity of the car is stopped in a distance of 3 feet. Then $H = \frac{V^2}{64.32} = \frac{88 \times 88}{64.32} = 120$ ft equivalent drop to produce this speed. Then the energy developed would be

$$3,600 \text{ lb} \times 120 \text{ ft} = 432,000 \text{ ft lb}$$

at the instant of contact. Then

$$\frac{432,000 \text{ ft lb}}{3 \text{ ft stopping distance}} = 143,500 \text{ lb}$$

or nearly 72 tons impact. These examples should amply demonstrate the urgent need for applying a load, or taking up a slack cable, slowly.

CHAPTER II

FIBER ROPE

For thousands of years man has made use of vegetable materials for hauling and lifting loads. At first, vine stems were twisted together to form a crude strand that had sufficient strength for his requirements, yet was flexible and could be tied into knots. Use had been made at various times of such materials as the fibrous bark of the palm tree, the hair of the coconut, camel hair, horse hair, thongs cut from hides, cotton, jute, sisal, flax, and wild and cultivated hemp, until today we have the modern manila rope.

Evolution has passed even the manila rope and produced the wire rope or steel cable as it is commonly known. While wire rope may properly be considered an improvement over manila rope, it has not entirely superseded manila rope, for each has its own field of use. Wire rope is used for permanent installations and where heavy loads are to be lifted. Manila rope, on the other hand, is generally used on temporary work such as construction and painting jobs and in marine work.

There is a great difference between wire rope and manila rope, not only in strength but in the very nature of the material itself. Steel wire is produced synthetically to meet the exacting specifications of the metallurgist. Since it has a known composition, the strength, modulus of elasticity, and other properties are all definitely known even before the rope is manufactured, and all wire ropes of the same grade of steel and of the same type of construction have the same physical properties.

Manila fibers, on the other hand, are a product of Nature, being obtained from the abaca, or wild banana plant, grown in the Philippines, and, as such, vary in quality, their strength being dependent upon climatic conditions, fertility of the soil, cultivation, and curing.

The fibers taken from the outer portion of the stalk of the abaca plant are somewhat darker than those taken from nearer the center, and are slightly lower in strength and less durable. The

14

fibers are carefully sorted into grades before being imported by the rope manufacturer. The individual fibers as received are from 6 to 15 ft long, but the hanks must be cut to about 7 ft in length in order to pass through the drawing frames, which comb out the fibers parallel to each other. To lubricate and preserve the rope, cordage or special oils are applied to the fibers during the combing process.

The fibers, or slivers, come out in a fluffy stream and are then recombed six or seven times. During this process combed fibers from other shipments of the same grade are blended with it to ensure a uniform quality. Each time the fibers are combed, their number becomes less, until the sliver is reduced to the proper size to be spun into a yarn or thread, which is the unit of rope construction. A number of yarns are then twisted left-handed into a strand, and three or four strands are laid right-handed into the finished rope.

Material. Manila rope is recommended for most hoisting operations calling for the use of fiber rope. During the war, when manila was not procurable, substitutes such as sisal (pronounced "*sy*-sal"), jute, hemp, and cotton had to be used. The materials and grades of natural-fiber ropes most commonly encountered are as follows:

1. Yacht rope—The highest quality manila of very fine appearance. Costly, but used on special jobs where appearance is an important factor.
2. Bolt rope—High-grade manila rope, about 10 to 15 per cent stronger than No. 1 grade manila.
3. No. 1 manila rope—The standard high-grade hoisting rope used for important rigging jobs. Usually identified by a trade marker.
4. No. 2 manila rope—Has about the same initial strength as No. 1 grade, but loses its strength more rapidly in service.
5. No. 3 manila rope—Has about the same initial strength as No. 1 grade, but loses its strength very rapidly in service.
6. Hardware-store rope—A very poor grade of manila rope, of low strength and short life.
7. American hemp rope (tarred)—Has about 80 per cent of the strength of No. 1 manila rope.
8. Java sisal rope—Has about 75 per cent of the strength of No. 1 manila rope.

TABLE I. PROPERTIES OF FIBER AND SYNTHETIC ROPES

Properties	Manila 1st Grade	Filament Nylon	Filament Dacron	Polyethylene	Polypropylene	Polyamid	Glass	Improved Plow Steel 6 × 19
Safety Factor Used	10	5	5	5	5			5
Strength, dry (Safe Working Loads)								
¼ in. dia.	60#	360#	240#	210#	250#			1,096
	135	800	520	420	540			2,440
	265	1,460	900	720	900			4,280
	440	2,180	1,380	1,080	1,320			6,780
	540	3,120	1,900	1,400	1,600			9,520
	770	4,280	2,520	1,900	2,150			12,870
	900	5,400	3,300	2,400	2,500			16,740
3 in. Circ.	1,200	6,700	4,200		3,400			(For comparison only)
3½ in.	1,500	9,300	5,600		4,400			
4½	1,850	11,300	7,000		5,600			
5½	2,250	14,000	8,000		7,000			
	2,650	16,650	10,000		8,400			
6	3,100	19,600	12,000		10,000			
Strength, wet Dry strength = 100	100+	85–90	100	105	105	90–95	50–75	
Weight, dry Dry manila rope = 100	100	84	104	66	63			523
Flexing, Resistance to repeated	F	G	G	G	E	G	P	
Abrasion, Resistance to	F	F	G	G	G	G	P	
Acids, Resistance to	P	G	G	E	G		E	
Alkalis, Resistance to		*E	E	E	G		E	
Oils, Resistance to	G	P	E	P			E	
Sun light, Resistance to	P	E	E	E	G			
Solvents, Resistance to		E	E	E			E	
Mildew, decay, organisms, Resistance to	F	F		G	E	G	E	
High Temperature, Use at	F	G		**G	G		P	
Low Temperature, Use at	F	H	A+	H	F	H		
Wet, Use when	A			E	E		E	
Stretch under load	G	G	G					
Dielectric strength, dry	N	Y	G	Y	F			
Splicing, Ease of	A	G		G				
Slippery to handle								
Work Life					E	E		

A = average   E = excellent   F = fair   G = good
H = high   P = poor   Y = yes   N = no
* = Except linseed oil   ** = Floats on water.

16

9. Henequen sisal rope—From Mexico and Cuba. Has about 60 per cent of the strength of No. 1 manila rope.
10. Jute rope—Has about 50 per cent of the strength of No. 1 manila rope.
11. Cotton rope—Has about 50 per cent of the strength of No. 1 manila rope.

Manila rope should be light yellow with a silvery or pearly luster, and should have a smooth waxy surface. No. 1 grade rope is very light in color, No. 2 grade is slightly darker, and No. 3 grade is considerably darker. Hardware-store rope is still darker, and the short fibers cause many fiber ends to protrude from the strands. Yacht and bolt rope, as might be expected, are very light in color but are distinguished from sisal which feels harsh and dry to the touch. Jute is rather dark brown and is very soft and flexible. Cotton, of course, is white and soft and should be readily recognized.

Synthetic-fiber ropes, such as nylon, dacron, polyethylene, and polypropylene are, to a large degree, replacing manila rope. These ropes have individual fibers running for their entire length, not short, overlapped fibers as in natural-fiber ropes, and consequently have greater strength. Synthetic ropes are also more resistant to decay, etc.

TABLE II. WEIGHTS OF MANILA ROPE

$\frac{1}{4}$ in.	.020# per ft
$\frac{3}{8}$	.041
$\frac{1}{2}$	.075
$\frac{5}{8}$	.115
$\frac{3}{4}$	.167
$\frac{7}{8}$	.205
3	.270
$3\frac{1}{2}$	.360
4	.480
$4\frac{1}{2}$	.600
5	.744
$5\frac{1}{2}$	.895
6	1.080

Purchasing. In purchasing manila rope for hoisting purposes, the specifications should require that it be made of first-grade manila fibers, and the order should be placed with a reputable manufacturer. Most rope manufacturers proudly place a marker of some kind in their first-grade rope for identification. Some have one or more yarns colored, or they may have a colored string in one strand, or they may have a tightly twisted paper ribbon with their trade-mark on it in one of the strands (Fig. 1). Rope without such a marker should not be used for hoisting purposes, as either it is not first-grade rope or it is made by a manufacturer who does not wish to be identified with it.

In ordering new rope the purchaser should specify the type of construction best suited to his requirements. Manila rope is usually made up with three strands. But for work where abrasion is a factor, a rope having four strands laid around a small rope core is recommended. This rope is more nearly round in cross section

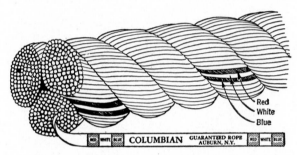

Red
White
Blue

FIG. 1. Three-strand manila rope of first grade identified by tape marker. (*Columbian Rope Co.*)

and is used quite extensively for power hoisting. Its strength, however, is very slightly less than three-strand rope.

In making a rope, the yarns can be formed into the strands and the strands can be laid into the rope either tightly or loosely, making what is known, respectively, as "hard-laid" or "soft-laid" rope. Hard-laid rope is stiffer and more resistant to abrasion, whereas soft-laid rope is limp but stronger. A medium lay is generally recommended.

The method of measuring manila rope causes confusion at times. Up to 1 in. diameter a rope is called by the diameter, but larger ropes are called by the circumference, which is assumed to be three (not $3\frac{1}{7}$) times the diameter.

Storage. Assuming that good manila rope has been procured, it is necessary to store it properly, both before using and between periods of use. If properly stored, it may remain in good condition for 3 or 4 years.

Rope should never be kept on the floor or in a box, closet, or small room where air circulation is restricted. Instead, it should be stored on a wood grating platform about 6 in. above the floor or hung up in loose coils on large-diameter wooden pegs. (A discarded 12-qt pail nailed to the wall makes a good bracket for hanging rope.)

Rope should be protected from the weather and from dampness from any cause. A wet rope should be thoroughly dried and cleaned before being stored away; otherwise it will lose its strength very rapidly. On the other hand, it should be kept away from boilers, radiators, steam pipes, and other sources of heat. In fact, the ultraviolet in the sun's rays is detrimental to

FIG. 2. The rope unwinds counterclockwise as it comes out of the coil. (*Columbian Rope Co.*)

manila rope. A temperature of 50 to 70°F and a humidity of 40 to 60 per cent are recommended.

Exposure to stack gases or to carbon monoxide or carbon dioxide will cause rapid deterioration of a rope.

Opening Up a New Coil. In opening up a coil of new rope, the instructions printed on the tag should be carefully followed. After the burlap wrapping has been opened or removed, look inside the coil for the end of the rope. This should be at the bottom of the core, or eye. If it is not, turn the coil over so that the end will be at the bottom. Cut the lashings that bind the coil together, and pull the end of the rope up through the core. As the rope comes out of the coil, it will then unwind in a counterclockwise direction (Fig. 2). Even though the rope is unwound properly, loops or hockles may form in it, and these should be carefully removed.   If the rope is unwound improperly, you will be well aware of the fact, as the hockles will develop into numerous kinks, which, if pulled out, will cause severe damage to the rope.

Should circumstances require the uncoiling from the outside of the coil, place the coil so that, as the rope pays off, it will unwind in a counterclockwise direction.   If a rope should have a large number of kinks, coil it on the floor counterclockwise, then pass the end through the coil, and proceed to uncoil.   This should remove the kinks.

When a rope is cut to the required length, whippings of yarn should be applied at once to the ends to prevent the strands from becoming unlayed.   Otherwise, the strands may slip in relation to each other, causing one of them to assume more or less than its share of the load, which will result in a shortened life of the rope.

Strength of Rope.   Table I gives the working strength of a three-strand rope of various materials when subjected to direct pull.   A factor of safety of 10 has been used for natural-fiber ropes; 5 for synthetic-fiber ropes.   (The "factor of safety" is the breaking strength divided by the safe working load.)   If fiber rope is used for high-speed power hoisting, a higher factor of safety should be used.

Handling Rope.   When a rope is attached to a hook, ring, or pulley block, a thimble should be placed in the loop or eye to reduce the wear on the rope and to decrease the stresses developed in the rope when it is bent around a very small diameter.   When a rope is used as a sling, it should not pass over the sharp edges of castings, boxes, or beams unless a padding is used.   Also consideration should be given to the angle between the legs of a sling when placed in position to pick up a load.   If the ropes make an angle to the horizontal of 60 deg, they have only 86 per cent of their strength when vertical.   At 45 deg their strength is 71 per cent, and at 30 deg it is only 50 per cent, this not including the loss of strength due to sharp bending.

Rope should not be dragged over concrete pavement or through sand or cinders, as the outside surface of the rope will be worn and cut by the abrasive action.   Also, small particles of gritty material may get between the component parts of the rope and cause internal damage.

Moisture.   As far as is practicable, rope should not be allowed to become wet, as this not only hastens decay but also causes the rope to kink very readily.   Fiber ropes used between fixed objects, such as in the case of guys on a gin pole, should be slackened when it begins to rain; otherwise shrinkage may over-strain the wet rope and damage it.

Should a wet rope become frozen, it must not be disturbed until completely thawed; otherwise the frozen fibers will be broken as they resist bending. Although every effort should be made to prevent rope from freezing, little harm will be done if the above instruction is followed.

While not recommended for general practice, it should be borne in mind that a rope splice which has been soaked in water will be stronger than the dry body of the rope. Likewise, a wet rope is about 10 per cent stronger than a dry rope.

Chemicals. Rope should be kept away from acids and other destructive chemicals, as exposure to even the fumes will materially weaken it.

Reeving. Each time a rope passes over a sheave, friction is produced, and the pull exerted on the object is proportionately

TABLE III. SAFE LOADS FOR MANILA-ROPE FALLS

(Using factor of safety of 10.)

Rope diameter, in.	Parts of rope supporting load, lb					
	1 part	2 parts	3 parts	4 parts	5 parts	6 parts
$\frac{1}{2}$	240	440	600	725		
$\frac{3}{4}$	490	890	1,200	1,480	1,680	1,830
1	820	1,490	2,030	2,460	2,800	3,050
$1\frac{1}{4}$	1,230	2,230	3,050	3,700	4,200	4,580
$1\frac{1}{2}$	1,680	3,060	4,180	5,070	5,750	6,280

less than the pull exerted by the source of power. When the load is at rest, the tension on the rope is equal to the load divided by the number of parts of the rope actually supporting it. For instance, if the lower block has two sheaves, there are four parts of rope. If it has two sheaves and the rope is dead-ended at the lower block, then there are five parts of rope supporting the load.

As soon as hoisting is begun, friction causes an additional load on the rope of about 10 per cent for each sheave passed over. This is illustrated in Table III.

Factor of Safety. One of the principal advantages of rope of any description is the ability to lift heavy loads with a reasonable pull on the hauling line. This is accomplished by means of pulley blocks.

In general, the load to be lifted is divided by the number of parts of rope supporting the lower pulley block to obtain the pull on the hauling line.   This, however, is true only with the load in a static or stationary condition.   The actual strain on the hauling

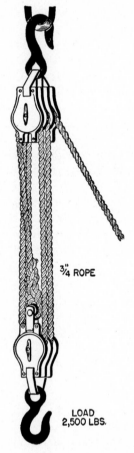

¾" ROPE

LOAD
2,500 LBS.

Fig. 3.   A five-part rope fall.

line, when hoisting, is somewhat greater than this figure owing to friction of the pulleys, as shown in Table III.

A new manila rope ¾ in. in diameter has a nominal breaking strength of about 5,400 lb, according to the rope manufacturers, but this does not mean that 5,400 lb can be lifted on it.   For ordinary slow-speed hoisting where men's lives or valuable equipment are involved, many safety engineers believe that the safe load should not exceed one-tenth of the breaking strength, or 540 lb. In other words, a factor of safety of 10 should be maintained.

This wide margin of 540 to 5,400 lb may appear to be excessive. However, there are many indeterminate factors that must be taken into consideration.

Let us emphasize the fact that errors in judgment are not always cumulative, although it is very possible for this to happen. We shall consider a hypothetical case. Let us assume that it is desired to lift a weight thought to be about 2,500 lb on a five-part $\frac{3}{4}$-in. manila rope fall, that is, using a double and triple pulley block as shown in Fig. 3. In order to obtain the static tension on the rope, the weight (2,500 lb) is divided by the number of parts of rope supporting that weight (five), which gives a load of 500 lb. This is only 8 per cent less than the allowable safe load of 540 lb and may be acceptable.

However, the load to be lifted has only been *estimated* to weigh 2,500 lb; it may be more or it may be less. Very few men can estimate the weight of an object within 25 per cent unless they calculate it mathematically, so we shall assume that the actual weight happens to be 10 per cent heavier than estimated, or 2,750 lb. This means that the actual static rope tension is also 10 per cent higher than our first figure, or 550 lb.

Now, each time a rope passes over a sheave in a pulley block, friction is developed. For ordinary calculations it is usually taken as 10 per cent of the rope tension for each sheave passed over. There being five such sheaves, the pull on the hauling line when the load is being lifted uniformly is about 0.32 times 2,750 lb, or 880 lb.

But few loads can be lifted smoothly. They may swing, or they may vibrate owing to worn winch gears, or they may be accelerated rapidly, or in the case of hand-power hoisting the pull may be jerky. This greatly increases the stress in the rope, sometimes several hundred per cent. In our imaginary problem let us assume that the stress is thus increased 100 per cent, which is not unreasonable. This gives a true tension on the hauling rope under actual working conditions of twice 880 lb, or 1,760 lb.

Now let us turn our attention to the rope itself. When the rope was new, its nominal breaking strength was 5,400 lb. This was the strength of the rope when brand new. Our rope, however, has seen some hard service. Fibers and even a few yarns are broken. The rope has somewhat dried out, and it is otherwise reduced in strength, although it is still serviceable. In other words, it is the type of rope that is frequently used for such purposes. Let us assume that it has lost 25 per cent of its original strength, or that the present strength is about 4,050 lb when subjected to a straight pull.

Now it is a characteristic of any rope that it cannot be used without being bent. In bending a rope over a pulley or around the eye of a hook or around itself as in a knot, the individual fibers are subjected to severe bending stress, which must be deducted from the breaking strength of the rope to obtain the net strength.

Tests show that sharp bends and knots frequently reduce the strength of a rope as much as 50 per cent, so we shall assume that our hypothetical rope has been thus weakened and the net strength reduced to about 2,025 lb. This is the maximum load which the rope may be expected to carry.

We started our calculations by believing that we had a factor of safety of 10, or that the rope had a strength ten times as great as the load imposed upon it. Actually, we find that the strength is about 2,025 lb while the load on it is about 1,760 lb, which gives a true safety factor of only 1.15. Keep in mind that a factor of safety of 1 means no safety at all and that the rope will fail.

Again, let it be said that this may be an exceptional but by no means impossible combination of conditions. This analysis, it is hoped, has demonstrated why a factor of safety of 10 is necessary when only the estimated weight and the nominal strength of a rope are known.

Many rope users and, in fact, some rope manufacturers consider a safety factor of 5 or even lower sufficient, but in the interest of safety to personnel and to property it is believed that a factor of safety of 10 is absolutely necessary.

Lay of the Rope. Practically all fiber ropes are laid up right-handed, so only right-lay rope has been considered in this discussion. Soft-laid rope is slightly stronger than hard-laid rope but is less resistant to abrasion. For general use a medium lay is recommended.

Waterproofing. Most rope manufacturers produce a special waterproofed rope bearing a trade name. Waterproofing helps a rope to retain its flexibility even though wet, reduces the liability of kinking, and makes the rope easier to handle. All this is, of course, in addition to preserving the rope from decay.

Inspection. In order intelligently to estimate the loss of strength in a worn rope, such as was just described, it should be inspected periodically.

When a used rope is inspected, the first touch will frequently give an idea of its general condition, but such a casual examination is not sufficient. A rope upon which a man expects to trust his life should not be inspected in any but the most thorough manner. This means going over every foot of its length. Like the proverbial chain, a rope is only as strong as its weakest part, and it is the inspector's duty to find that part.

Examine the outside of the rope, and observe the number of broken fibers or broken yarns, keeping in mind that these failures represent just so much loss of strength. Do not be misled by dirt on the surface; any rope that has been used will be dirty on the outside. Open up the rope by untwisting the strands so that you can observe the condition of the inside, but try not to kink the strands. The interior of the rope should be as bright and clean as when the rope was new. Observe if there are any broken yarns inside or if there is an accumulation of a powderlike sawdust, which indicates excessive internal wear between the strands as the rope is flexed back and forth in use.

Then, if the rope is large enough to permit, open up a strand and with a lead pencil or other blunt instrument try to pull out one of the *inside* yarns, keeping in mind that, if a rope has been overloaded, it is the interior yarns that will have failed first. Excessive oil on the outside of a new rope is also an indication that it has been overloaded.

If it is four-strand rope with a heart, try gently to pull out the heart. If the heart readily comes out in short pieces, this rope likewise has been overloaded and should not be used for hoisting.

If possible, pull out a couple of long fibers from the end of the rope and try to break them between the hands. The finer fibers are relatively stronger than the coarser ones, and all should be broken only with difficulty. In fact, believe it or not, some fibers have a tensile strength as high as 30,000 lb per sq in.

If the inside of the rope is dirty, if the strands have begun to unlay, or if the rope has lost its "life" and elasticity, it should not be used for hoisting purposes. If the rope is high-stranded and presents a spiral appearance, or if the heart protrudes, the load will not be equally distributed on the strands and a very short life may be expected.

Often the surface of a rope feels dry and brittle, or it may show evidence of having been in contact with a hot pipe or other source of heat, or it may be discolored as the result of exposure to acid fumes, in which cases it should be discarded.

Condemning an Unsafe Rope. When a rope has been condemned, it should be destroyed at once or cut up into short hand lines so that it cannot again be used for hoisting purposes. Some organizations even prohibit the use of rope over 2 years old for scaffold falls. To ensure that older rope will not be used, their pulley blocks are painted a different color each time a new rope is installed, in order to indicate which are new and which are last year's ropes. Also the date of reroping is stamped on the lower or single block. Judging the condition and strength of a manila rope is not so simple as you may have been led to believe. It takes a lot of experience and good judgment, and even then the chances of error are very great.

TABLE IV. ELECTRICAL CONDUCTIVITY OF MANILA ROPE

(5 ft between electrodes)

Rope	Exposure to rain, hr	Current leakage observed at, volts	Severe leakage at, volts
Ordinary..................	Dry (no exposure)	160,000	
Waterproofed.............	Dry (no exposure)	160,000	
Ordinary..................	$2\frac{1}{2}$	8,000	11,000
Waterproofed.............	$2\frac{1}{2}$	21,000	25,500
Ordinary..................	4	5,000	7,000
Waterproofed.............	4	17,000	19,000

General Notes. Manila rope, when dry, is an excellent electrical insulator. The dielectric strength or insulating ability of a wet rope depends upon the percentage of moisture in the rope. Inasmuch as waterproofed rope absorbs less moisture than ordinary rope, it is a better insulator and therefore is safer for use when material must be hoisted in the vicinity of live electrical apparatus. Data in Table IV were furnished by Columbian Rope Company and are the result of tests made on 5-ft-long samples of manila rope suspended vertically with a 6-lb weight attached to the lower end.

Manila rope which is loaded to 50 per cent of its breaking strength (a safety factor of 2) may be expected to fail within a few hours; if loaded to 75 per cent it may fail within a few minutes. This is due to the creepage of the fibers on each other.

Synthetic ropes have rather slippery fibers. Therefore, all splices in such ropes should have at least five tucks. Braided sash cord is not intended for hoisting. This cotton cord, however, is frequently used for signal cords, etc. The breaking strength is about 320 lb for $\frac{1}{4}$-in. and 780 lb for $\frac{3}{8}$-in. sash cord.

Knots, Bends, and Hitches. There are hundreds, if not thousands, of different rope knots, bends, and hitches, but the average rigger can get along with the knowledge of a comparatively few. Technically speaking a "knot" is the intertwining of the end of a rope with a portion of the rope. A "bend" is the intertwining of the ends of two ropes to make one continuous rope. A "hitch" is the attachment of a rope to a post, pole, ring, hook, or other object. A "splice" is the joining of the ends of two ropes or of the end of a rope with the body of the rope by weaving the strands over and under the strands of the other part. Actually, the names are not always correct designations; for instance, the bucket hitch is also known as the fisherman's bend or the anchor knot. A few of the most important fastenings are shown in Figs. 4 to 108, inclusive, those marked by a star ($\star$) being the most commonly used.

There is one little trick to remember in tying any knot if you do not want it to slip; if two parts of the rope are squeezed together, they should be arranged so as to have a tendency to pull in *opposite* directions. For instance, compare the reef knot with the granny.

FIG. 4. Bight, or simple loop. A loop in a rope, as distinguished from its ends. A part of all knots.

FIG. 5. Overhand knot, or single knot. To prevent unreeving. Forms eccentrically on rope. Part of many knots. Efficiency 45 per cent of strength of straight rope.

FIG. 6. Figure-eight knot. To prevent unreeving.

FIG. 7. Marline spike hitch, or boat knot. For drawing seizing tight.

FIG. 8. Reef knot, square knot, sailor's knot, or flat knot.★ For joining two ropes of same size, but not safe for ropes of different sizes. Will not hold if ropes are wet. Efficiency 50 per cent.

UNSAFE

FIG. 9. Granny knot, or lubber's knot. Improperly made reef knot. May slip or jam. *Unsafe.*

UNSAFE

FIG. 10. Thief knot. Improperly made reef knot. Will slip. *Unsafe.*

FIG. 11. Sheet bend, becket bend, hawser bend, or weaver's knot.★ Shown as used for joining ropes of different sizes. More secure than reef knot. Difficult to untie.

FIG. 12. Sheet bend in eye splice. For securing rope to permanent eye or loop.

FIG. 13. Double becket hitch. To secure rope to permanent eye or loop. More secure than sheet bend when used in spliced eye.

FIG. 14. Fisherman's knot. Not dependable except in very flexible twine or cord. For use with cordage of different sizes.

FIG. 15. Overhand knot, or rosette knot. For joining two ropes. Will not slip but is of low efficiency.

FIG. 16. Hawser bend. For joining large hawsers.

FIG. 17. Reeving-line bend. For joining ropes of different sizes.

FIG. 18. Surgeon's knot. For tying up a compressible bundle, such as canvas.

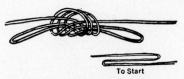

To Start

FIG. 19. Knot shortening. For shortening a rope without cutting it. An overhand knot using all three parts of the rope. Knot cannot be tied unless one end of the rope is free.

FIG. 20. Sheepshank.★ For shortening a rope without cutting it.

FIG. 21. Sheepshank cut. A knot of average strength when under a steady pull but one that will part if slackened and jerked. Made by forming sheepshank and cutting the center part.

FIG. 22. Bowline.★ A nonslipping eye. Will not jam. Easily untied. Efficiency 60 per cent.

FIG. 23. Bowline on a bight. A nonslipping eye. Begin as an ordinary bowline, using two parts of the rope. After passing bight through loop, open it up and slip it over the entire loop.

FIG. 24. Double bowline. A nonslipping eye. A comfortable sling to sit in.

FIG. 25. Running bowline. A strong running loop or noose.

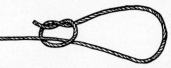

FIG. 26. Slip knot, or running knot. Not safe for heavy loads. Sometimes made with overhand knot at short end of rope.

FIG. 27. Midshipman's knot. An adjustable noose.

FIG. 28. Fisherman's eye knot, Flemish loop. A fixed loop. Used on small-size cord, etc.

FIG. 29. Double overhand knot, or loop knot. A fixed eye. Lacks strength. Will not slip, but jams.

FIG. 30. Two half hitches, or studding-sail tack bend.★ For attaching rope to a ring.

FIG. 35. Studding-sail hitch. For fastening a rope at right angle to a post. For hoisting timber, etc.

FIG. 31. Round turn and double hitch. For securing a rope to a ring. Should not be used for hoisting.

FIG. 32. Slippery ring. For temporarily securing a rope to a ring. Easily untied by pulling end of rope.

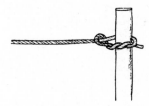

FIG. 36. Timber hitch. For fastening a rope at right angle to a post. Will not slip under load, but will readily loosen when strain is relieved. Efficiency 65 per cent.

FIG. 33. Half hitch and seizing. Not recommended unless end of rope is strongly seized to standing part.

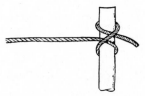

FIG. 37. Clove hitch, two half hitches, heaving line bend, or builders' knot.★ For securing a rope at right angle to a post.

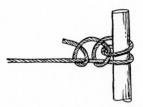

FIG. 34. Fisherman's bend, anchor knot, or bucket hitch. For fastening a rope to a ring or post.

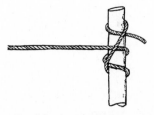

FIG. 38. Mooring knot, or magnus hitch (also see rolling hitch). For fastening a rope at right angle to a post.

Fig. 39. Mousing.★ Rope yarn placed on hook to prevent load from becoming detached.

Fig. 44. Sheet bend, hawser bend, weaver's knot, mesh knot, or netting knot. Method of joining ropes or cords in a net.

Fig. 40. Blackwall hitch.★ For temporarily attaching a rope to a hook. Will not slip if rope is dry.

Fig. 45. 90-deg net splice. For joining ropes of a net that run at 90-deg angle to each other. Unlay one strand of one of the ropes, tuck this strand under one strand of the other rope, and then re-lay in its original position in the rope.

Fig. 41. Midshipman's hitch. For temporarily attaching a rope to a hook. Will hold even if rope is wet.

Fig. 42. Double Blackwall hitch. For temporarily attaching a rope to a hook. Will not slip even if rope is wet or greasy.

Fig. 46. 60-deg net splice. For joining ropes of a net that run at 60-deg angle to each other. With free end of ropes toward top of sketch, open up the white rope and pass shaded rope under two strands of the former. Then open up the shaded rope and pass the end of the white rope under two strands of the shaded rope. Then pull together as shown.

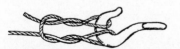

Fig. 43. Cat's-paw, or rocking hitch. For attaching a rope to a tackle block.

FIG. 47. Strap on a rope. A means of securing a grip on a rope under strain.

FIG. 52. Telegraph hitch. For hauling poles and logs.

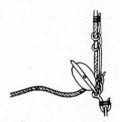

FIG. 48. Strap on a rope. A means of securing a grip on a rope temporarily so as to relieve the strain on the latter.

FIG. 53. Wall knot. A stopper, to prevent unlaying of strands at end of rope. Generally used in combination with crown knot.

FIG. 49. Rolling hitch.★ For fastening a rope to another rope or a pole parallel to it. For attaching a life-belt rope to a hanging life line. Grips tightly if pulled in direction shown, but is easily moved along rope or pole when strain is relieved.

FIG. 54. Crown knot.★ A stopper, to prevent unlaying of strands at end of rope. Generally used in combination with wall knot.

FIG. 55. Stevedore's knot. To prevent unreeving of rope.

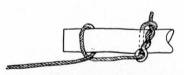

FIG. 50. Timber hitch and half hitch.★ For hoisting or hauling poles, pipes, etc. Easily released when strain is relieved.

FIG. 51. Well pipe hitch. For hoisting or hauling pipes and poles.

FIG. 56. Harness hitch. For attaching men to a toe line, loops being placed over the men's bodies. If only one end of rope is subject to pull, knot will slip.

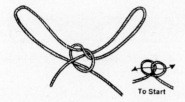

FIG. 57. Single pitcher knot, or rope handcuffs. A sling for handling balls, round rocks, etc. Also loops may be used as handcuffs.

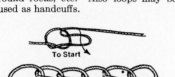

FIG. 58. Chain knot. For shortening a rope without cutting it. For forming a hand grip on a rope.

FIG. 59. Marling hitch. For bundling up canvas, etc.

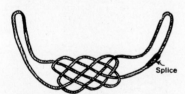

FIG. 60. Western ocean plant. A sling for handling shafts, armatures, etc.

FIG. 61. Whipping.★ To prevent unlaying of strands at end of rope. For binding a split pole or for locking two poles together. Start by placing bight along rope, then wind cord on rope starting from bight end *A*. Finish by passing end of cord *B* through bight, which is under whipping. Then pull on end *A* until interlocked bights are under middle of whipping. Cut off loose ends.

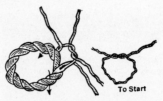

FIG. 62. Grommet. An endless rope of one strand only, it makes three complete loops. To terminate, divide strands as shown and tuck as in short splice.

FIG. 63. Eye splice.★ Tuck each strand four times. Then, if desired, divide each strand in half and cut off one end. Tuck the other, and then divide; cut, and tuck again to make a neater job.

FIG. 64. Flemish eye splice. Unlay one strand the required distance. Then form a loop of the two remaining strands, and lay the first strand back in its former place in the rope but in the reversed position relative to the other strands. Terminate by tucking as in ordinary eye splice.

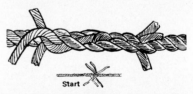

FIG. 65. Short splice.★ For joining two ropes together. Splice will not pass through pulley blocks. Unlay each rope fifteen times its diameter. Place ropes together as shown. Tuck each strand under one strand in the other rope, then over one strand and under one strand. This makes two tucks each side of center.

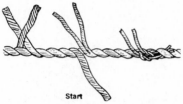

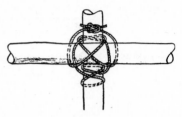

Fig. 66.   Long splice.   For joining two ropes together (three-strand rope shown). Will pass through pulley blocks.   Unlay each rope for a distance equal to twenty-five times its diameter, and place together as shown.   Unlay one strand in the left-hand rope, and wind in its place one strand from the right-hand rope. Repeat on other rope.   To terminate, divide adjacent strands in half, tie half strands of left-hand rope to half strands of right-hand rope, and then tuck ends under strands.

Fig. 69.   Pole and ledger lashing.   For lashing poles at right angles to each other.

Fig. 67.   Cut splice.   An eye in the middle of a continuous rope.   Place ropes as shown, and tuck as in eye splice.

Fig. 70.   Pole lashing.   For lashing two poles   together.   Two   such   lashings should be used.

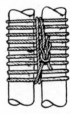

Fig. 71.   Portuguese knot, or necklace tie.   For lashing shear legs together. Spreading of legs puts strain on knot.

Fig. 68.   Flagpole hitch.   Similar to a clove hitch.   For attaching a boatswain's chair to a pole.

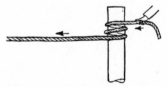

Fig. 72.   Snubber.★   For holding or slowly lowering a heavy load.   Strain on hand line is only a fraction of strain on load line.   Three or four turns should be taken.

FIG. 73. Spanish windlass.★ An endless rope used to pull two objects toward each other, similar to a turnbuckle. Stick is placed between parts of rope and rotated. To hold strain, the end of the stick is lashed to a fixed object.

FIG. 77. Scaffold hitch. For suspending a scaffold plank. Will prevent plank from tilting. Can also be used on plank on edge.

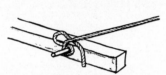

FIG. 74. Slippery hitch. A rope anchorage that is readily released by pulling on short end of rope.

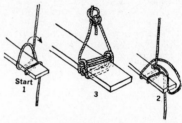

FIG. 78. Scaffold hitch. For suspending a scaffold plank. Will prevent plank from tilting. Can also be used on plank on edge.

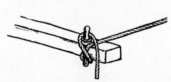

FIG. 75. Belaying-pin hitch. An anchorage for securing a rope. Several figure-eight turns may be required to hold the load. Easily cast off.

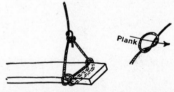

FIG. 79. Scaffold hitch. For suspending a scaffold plank. Will prevent plank from tilting. Can also be used on plank on edge.

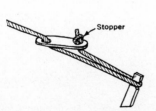

FIG. 76. Regulating lashing. For regulating the length of tent ropes.

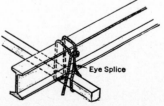

FIG. 80. Rivet scaffold hitch. For supporting needle beam of riveter's scaffold. Loose end of rope should be several feet long.

FIG. 81. Back hitch, or painter's scaffold hitch.★ For securing hoist rope to scaffold stirrup.

FIG. 82. Safety hitch for boatswain's chair. Make as shown, and slip loop under man's feet and up over his head. Cannot slip out of man's hands. To lower chair, feed rope into hitch.

FIG. 83. Barrel hitch, horizontal. For hoisting a barrel or similar object.

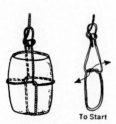

FIG. 84. Barrel hitch, vertical. For hoisting a barrel or similar object. Tie rope as shown, then pull parts as indicated by arrows.

FIG. 85. Sling for hoisting injured man. A cross lashing arranged as shown, and terminated with a bowline above the man's head.

FIG. 86. Back splice. A stopper to prevent unlaying of strands at end of rope. Also used as a handle. Start as a crown knot, then take several additional tucks.

FIG. 87. Crossed running knot. A slipknot.

FIG. 88. Reef knot on the bight. For shortening a rope without cutting. Both ends of rope may be anchored before tying knot.

FIG. 89. Buntline hitch. An adjustable noose.

Fig. 90. Double sheet bend. For joining ropes of different sizes.

Fig. 95. Double bowknot. For joining cords. Readily untied by pulling end of either cord.

Fig. 91. Lark's head with toggle. Used as a boat fastening. Released by removing toggle.

Fig. 96. Flemish knot. For joining two ropes.

Fig. 92. Carrick bend. For joining large hawsers. Will not jam.

Fig. 97. Shroud knot. A type of splice for joining two ropes. Place ropes together as in making short splice. Make wall knot in end of each rope about the other rope. Taper strands and serve rope.

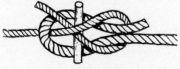

Fig. 93. Sheet bend with toggle. For joining ropes of different sizes. Will not jam.

Fig. 98. Single carrick bend. Similar to sheet bend. For joining two ropes of same or different sizes.

Fig. 94. Single bowknot. For joining cords. Readily untied by pulling on end of looped cord.

Fig. 99. Back-handed sailor's knot. For attaching a rope to a ring.

FIG. 100. Flemish eye knot. Safer than double overhand knot.

FIG. 105. Stopper knot. To prevent unlaying of strands.

FIG. 101. Killick hitch. For securing a rope to a post. Formerly used to attach rope to a stone anchor.

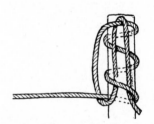

FIG. 106. Rope yarn knot. For splicing yarns used for seizing and other purposes.

FIG. 102. Diamond knot. A stopper to prevent unlaying of the strands at the end of a rope.

FIG. 103. Sheepshank, knotted. For shortening a rope without cutting it. More secure than ordinary sheepshank.

FIG. 107. Topsail halyard bend. For fastening a rope to a spar.

FIG. 104. Lanyard knot. To prevent unlaying of strands of a four-strand rope.

FIG. 108. Bell-ringer's knot. For temporarily tying up the lower end of a hanging rope.

FIG. 109. Coiling a rope, first position.

FIG. 110. Coiling a rope, second position.

FIG. 111. Coiling a rope, third position.

FIG. 112. Coiling a rope, fourth position.

Coiling a Rope. The rope should always be kept coiled up when not in use. An efficient method of making the coil is shown in the illustrations. Loop the rope over the left arm a number of times (Fig. 109) until about 15 ft of rope remains. In coiling the rope, careful watch should be kept for kinks, which of course should be removed. Then, starting about a foot from the top of the coil, the rope should be wrapped about six times around the loops by rolling them in the left hand (see Fig. 110).

Then the left hand is extended through the coil (Fig. 111) and the bight pulled back through the loops. Two half-hitches are tied around the bight, leaving a short end for carrying or for tying to a peg or supporting bar (Fig. 112).

CHAPTER III

WIRE ROPE

Wire rope is usually made up of a number of wires laid (not twisted) left-handed into a strand, then a number of strands, usually six or eight, laid right-handed around a hemp rope center to form the wire rope, or steel cable as it is frequently called. This

a. Right-lay, regular-lay rope.

b. Left-lay, regular-lay rope.

c, Right-lay, Lang-lay wire rope.

d. Left-lay, Lang-lay wire rope.

Fig. 1. Different lays of wire rope. (*American Chain & Cable Co.*)

makes what is called a "right-lay, regular-lay" rope (Fig. 1*a*). If the wires are laid right-handed and the strands left-handed, it would be called a "left-lay, regular-lay" rope. The great majority of wire ropes used for general hoisting purposes are of the former type.

Occasionally, both the wires in the strands and the strands in the rope are laid right-handed. Such a rope would be designated

41

as "right-lay, Lang-lay" rope, or if both are laid left-handed, it would be known as "left-lay, Lang-lay" (see Fig. 1c and d).

Construction. There are a great number of different possible constructions of wire rope. These are described by indicating first the number of strands, then the number of wires in each strand, namely, 6×7, 6×19, or 6×37. The smaller (and more numerous) the wires the more flexible the rope, but the less resistant to external abrasion. To overcome this, the Seale-lay rope was designed. It consists of comparatively small interior wires for flexibility with larger outside wires to resist abrasion. Warrington-lay rope has alternate larger and smaller outside wires. These facilitate its identification.

Figure 2 shows the exterior view as well as the cross section of some of the more common constructions. These views may help to identify a rope when its end cannot be examined, such as when it is spliced or socketed.

TABLE I. COMMON USES OF VARIOUS CONSTRUCTIONS OF WIRE ROPE

6×7 regular lay	Guys, highway guards, oil wells
6×7 Lang lay	Haulage systems having large sheaves and drums, cableways
6×19 Seale, regular lay	Mine incline hoists
6×19 Seale, Lang lay	Haulage systems, mine inclines, cableways
6×19 filler wire, regular lay	Miscellaneous hoists, derricks, cranes, tackle blocks, clamshell buckets, mine hoists, elevators
6×19 filler wire, Lang lay	Mine hoists, power shovels
6×19 Warrington, regular lay	Mine hoists, drum-type elevators, oil wells
6×37 regular lay	Heavy-duty cranes, mill hoists
6×37 Lang lay	Power shovels
8×19 Seale, regular lay	Hoists, traction elevators
8×19 filler wire, regular lay	Light-duty hoists, elevators
8×19 Warrington, regular lay	Light-duty hoists, elevators
Flattened strand	Coal unloaders, haulage lines, transmission
$6 \times 6 \times 7$ tiller rope	Small-boat steering, elevator hand ropes, signal ropes
$6 \times 6 \times 19$ cable lay	Heavy-duty cranes
18×7 nonrotating	Hoists having single-line suspension
Ropes with iron-wire-rope centers	On installations where cable overwinds on drums, where exposed to heat, or where additional strength is required

Ropes of 6 × 7 and 6 × 8 construction are too stiff to operate over sheaves and are therefore used only for guys, highway fences, etc. A very flexible rope is made of six 6 × 7 ropes laid into what is called "tiller" rope or 6 × 6 × 7 construction.

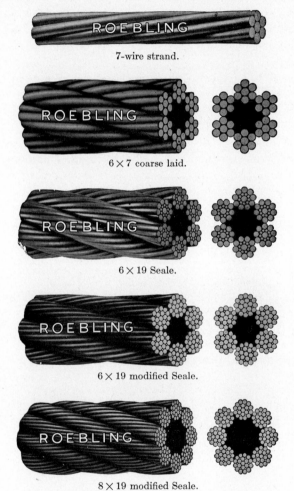

7-wire strand.

6 × 7 coarse laid.

6 × 19 Seale.

6 × 19 modified Seale.

8 × 19 modified Seale.

FIG. 2. Cross sections and exterior views of several constructions of wire rope. Exterior view may assist in identifying the type of construction when the end of the rope cannot be observed. (*John A. Roebling's Sons Co.*)

Most wire ropes are laid up with a hemp rope center to act as a cushion for the strands, but occasionally where the slight additional strength is needed or where the hemp center would be damaged by the heat, such as on foundry cranes, a wire-strand center or wire-rope center is used.

Where a load is suspended freely on a single cable and where the load would probably rotate owing to the strain on the rope, a special 18×7 construction is used. This is made up of a hemp center with 6 seven-wire strands laid left-handed and with 12 additional strands laid right-handed around them.

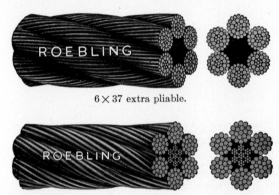

6 × 37 extra pliable.

6 × 19 modified Seale with independent wire-rope core, Lang-lay.

6 × 19 regular lay with independent wire-rope center.

18 × 7 nonrotating.

FIG. 2.   (*Continued*.)

For special applications triangular or oval strands can be used, thus presenting a more nearly round cross section and giving greater resistance to abrasion (see Table I for recommended construction).

Preforming. When the ordinary wire rope is laid up, the individual wires are under a slight bending strain, and hence when the rope is cut, the wires unlay and broom out if not held in place by seizings. In *preformed* rope any combination of wires and strands and any quality of steel can be used, but when laying up the wires into the strands and also the strands into the rope, they are bent slightly beyond the elastic limit in the curve they will finally assume, but they spring back sufficiently so that when in position they will lie "dead." An example of this preforming is the curve assumed by a piece of soft copper wire that has been pulled around a screw driver or hammer handle to remove the kinks. Although the desirability for seizing the end of wire rope is not eliminated, a preformed rope can be cut without the wires and strands unlaying.

Grades of Steel. Wire ropes are made of various grades of iron and steel, depending upon the needs.

Iron wire is low in strength (about 100,000 lb per sq in.) but is quite ductile and can be bent over small sheaves. It is sometimes used for elevators.

Traction steel is used principally for traction-drive elevators. It resists fatigue due to bending and causes a minimum of abrasion. Its strength is about 180,000 to 190,000 lb per sq in.

Mild plow steel is somewhat stronger (200,000 to 220,000 lb per sq in.) and is tougher; that is, it combines fatigue resistance with strength.

Plow steel is stronger and tougher than the mild grade. Its strength is about 220,000 to 240,000 lb per sq in.

Improved plow steel is the best grade, and each manufacturer usually applies a trade name to his rope. It possesses the highest strength (240,000 to 260,000 lb per sq in.) and abrasion-resistant properties.

Purchasing. In purchasing wire rope, it is essential that sufficient study be made to select the rope best suited to the job. The specification or purchase order should include the following information:

1. Length, such as . . . .   240 ft
2. Size . . . . $\frac{3}{4}$ in.
3. Construction . . . .   6 × 19 filler wire
4. Right or left lay . . . .   Right lay
5. Regular or Lang lay . . . .   Regular lay
6. Preformed or not . . . .   Preformed
7. Grade of steel . . . .   Improved plow steel

Identification. Many wire-rope manufacturers identify their *improved plow steel* rope by a trade name and by a colored strand or other means, such as

Roebling...................	"Blue Center"
Leschen....................	"Hercules"
American Steel and Wire.......	"Monitor" and "Tico Special"
Macwhyte.................	"Monarch Whyte Strand"
Broderick & Bascom.........	"Yellow Strand"
Bethlehem Steel.............	"Purple Strand"
American Chain and Cable....	"Green Strand"

Storage. Wire rope should be stored in a cool dry place away from fumes, chemicals, local heat, dampness, etc. It should be kept on the reel or spool until used. If it is to be stored for a long period of time, the outer layer of rope should be coated with a good moisture-resistant lubricant. Tar or asphalt in any form should not be used.

Fig. 3.   The proper method of removing wire rope from a reel mounted on trunnions.   (*American Chain & Cable Co.*)

Taking from Stock. When a length of wire rope is removed from its spool, it is important that the spool be rotated, either on a spindle resting on trunnions (Fig. 3), on a small turntable (Fig. 4)

FIG. 4. The proper method of removing wire rope from a reel mounted on a turn-table (*American Chain & Cable Co.*)

FIG. 5. The proper method of removing wire rope from a coil by rolling it along the floor. (*American Chain & Cable Co.*)

or by rolling the spool or coil along the floor as the rope pays out (Fig. 5). The rope should never be taken off one side of a spool or coil, as a kink will be produced for each wrap on the spool.

If it is absolutely impossible, because of space limitations or otherwise, to remove the rope from the spool by one of the methods mentioned above, then it may be necessary to resort to the following procedure. With the spool resting on its flanges, unwind several wraps of rope to accumulate sufficient slack. Then back up the rope to make a loose loop of rope on the spool, slip one loop off the right flange, and allow this loop to lie on the floor. Then slip a similar loop off the left flange, and let it also

Fig. 6. A means of removing wire rope from a reel that cannot be rotated on trunnions or on a turntable.

lie on the floor. (The rope on the floor will be in a figure eight. See Fig. 6.) Repeat this procedure, first on the right and then on the left, until the required length of rope has been unreeled. Then roll the spool back off the accumulation of rope on the floor, and pull away the end of the rope. The rope should come out of the pile without kinks.

Incidentally, if it is desired to roughly estimate the length of rope on a reel or spool the formula shown in Fig. 7 may prove very handy.

Kinking. Keep in mind that, once a kink has been produced in a rope, no amount of twisting or strain can completely remove it and the rope is weakened and may be unsafe for use. A slight, sharp, angular bend, known as a "dog leg," resulting from a partial kink, will chafe on the flanges of each sheave it passes over, and the rope will be prematurely worn at that spot.

Seizing. Before cutting any wire rope (including preformed rope), apply three seizings each side of the location of the proposed cut, then cut the rope. Do not cut first and then seize. If cutting is done by means of an acetylene torch, sometimes the seizings are omitted, as the wires and strands are simultaneously welded together.

Measuring. All ropes, both wire and manila, are measured by the diameter of the circle that will enclose the cross section of the

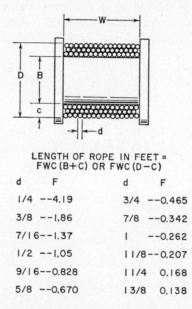

LENGTH OF ROPE IN FEET =
FWC (B+C) OR FWC (D−C)

d	F	d	F
1/4	--4.19	3/4	--0.465
3/8	--1.86	7/8	--0.342
7/16	--1.37	1	--0.262
1/2	--1.05	1 1/8	--0.207
9/16	--0.828	1 1/4	0.168
5/8	--0.670	1 3/8	0.138

APPROXIMATE LENGTH OF ROPE
ON A DRUM OR REEL

FIG. 7. Formula for making a rough estimate of the length of wire rope on a spool or reel. (*Bethlehem Steel Co.*)

rope (see Fig. 8). In other words, the calipers or caliper rule should be rotated or moved longitudinally until the largest diameter is found.

All new wire ropes are made slightly oversize, and this may cause confusion at times.

Size of Rope, in.	Oversize, in.
$\frac{1}{4}$ to $\frac{3}{4}$	0 to $\frac{1}{32}$
$\frac{7}{8}$ to $1\frac{1}{8}$	0 to $\frac{3}{64}$
$1\frac{1}{4}$ to $1\frac{1}{2}$	0 to $\frac{1}{16}$

Lubrication.   When the wire rope is manufactured, lubricant is applied to the hemp center, which acts as a storage medium for the oil and pays it out as the rope is used.   Frequent application of proper lubricant to the exterior of a rope helps retain the original lubricant within the rope.

When lubricating a rope it is desirable to use a fairly viscous oil recommended by the rope manufacturer.   This oil should be heated and applied while quite "thin" so that it will get into the center of the rope, then cool and thicken, which will keep it from being thrown off by centrifugal force as the rope passes over the sheaves.

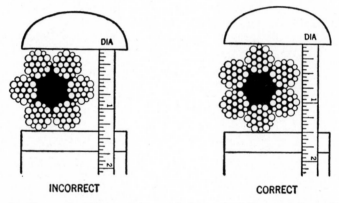

INCORRECT   CORRECT

Fɪɢ. 8.   The nominal diameter of a rope is the greatest diameter it is possible to measure.   (*John A. Roebling's Sons Co.*)

The best way to apply the heated lubricant is by causing a horizontal cable to move through a long pan of oil or by running a vertical cable through a can of oil with a tight-fitting bushing at the bottom.   Wipe off the surplus oil as the rope leaves the bath. Frequent application of proper lubricant during service is necessary to prevent the hemp core from becoming dry.   A dry-core rope will wear and crush more quickly than a lubricated one. Also it will absorb moisture more readily, which will result in interior corrosion of the rope.

On traction-drive elevators only a small quantity of very thip oil is used on the ropes in order not to reduce the adhesion between the ropes and the drum.   When any rope is observed with the lubricant dry and flaked, the rope should be cleaned with a wire brush and new lubricant applied.

Corrosion Corrosion or rust weakens a rope materially, yet it is very difficult, if not impossible, to estimate its weakening effect. A very rough idea can be obtained by carefully examining the exterior of the rope, and for lack of more definite data it must be assumed that the interior wires are in a like condition.

Chemicals. Wire rope must be kept away from all chemicals. Acid will attack the metal. Alkali may destroy the internal lubrication. Damage by chemicals is not always apparent and may greatly weaken a rope when least expected.

Design of Installation. It has been said of some machinery manufacturers that they spend much time and energy designing a

TABLE II. INSTALLATION OF CLIPS*

Rope diameter, in.	Number of Crosby clips for eye attachment	Spacing between clips, in.
$\frac{1}{4}$	2	$1\frac{1}{2}$
$\frac{5}{16}$	2	2
$\frac{3}{8}$	2	$2\frac{1}{4}$
$\frac{7}{16}$	2	$2\frac{1}{2}$
$\frac{1}{2}$	3	3
$\frac{5}{8}$	3	4
$\frac{3}{4}$	4	$4\frac{1}{2}$
$\frac{7}{8}$	4	$5\frac{1}{4}$
1	4	6
$1\frac{1}{8}$	5	7
$1\frac{1}{4}$	5	8

* The proper number and spacing of Crosby clips are important.

crane or hoist; then when it is completed, they place a wire rope on it. In other words, they give little or no thought to the rope. Anyway, if the rope does not last long, the rope manufacturer, rather than the hoist manufacturer, will probably be blamed by the operator.

A little thought given to the proper design of hoisting equipment should result in more satisfactory operation, longer life, and lower operating cost. Following are a number of suggestions to be considered in the design of wire-rope installations.

Attachments. Crosby clips are probably the most common method of attaching a wire rope to the equipment. The number of Crosby clips required is given in Table II. These clips should be spaced not less than six rope diameters apart, and *all* clips must be placed on the rope with the U bolts bearing upon the short or "dead" end of the rope (Fig. 9). Of course, a heavy-duty thimble should be provided for every eye, whether clipped or spliced. If properly made, a clipped eye should develop about

Fig. 9. Proper installation of Crosby clips. (1) Tighten clip farthest from eye. (2) Take a strain on the rope. (3) While the rope is under the strain, tighten the other clips. (*American Chain & Cable Co.*)

80 per cent of the strength of the rope. Beware of treacherous malleable-iron clips.

Bind the rope on itself at the toe of the thimble. Then apply the clip farthest from the thimble first, at about 4 in. from the end

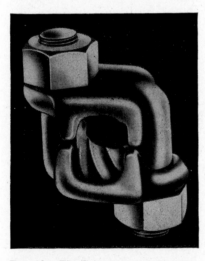

of the rope, and screw up tightly. Next, put on the clip nearest the thimble and apply the nuts handtight. Then put on the one or more intermediate clips handtight. Take a strain on the rope, and while the rope is under this strain, tighten all the clips previously left loose. In tightening the clips, take a turn alternately on the two nuts so as to keep the roddle of the clip square. After the rope has been in operation a short time, again tighten all the clips. No slack should exist in either part of the rope between the clips.

Double-base safety clips (Fig. 10) having corrugated

Fig. 10. Fist-Grip wire-rope clip develops greater strength and can be placed on the rope without crushing the strands.

jaws to fit both parts of the rope can be installed without regard as to which part bears on the live or dead parts of the rope. They ares aid to develop about 95 per cent of the strength of the rope; hence it is customary to use one clip less than indicated in the table for Crosby clips.

Several patented attachments are on the market, some of which can be recommended while others that develop high strength depend too much on the human element. If not attached to the rope according to specific directions, they may not be dependable. Each such clamp should be carefully investigated before adopting it for general use.

It is important, when using left-lay rope, that only clips or clamps designed for left-lay rope be used; otherwise the sharp ridges between the corrugations in the forging will run crossways rather than parallel to and between the strands of the rope. Cutting of the strands thus may result when using ordinary right-lay devices on left-lay rope. (Fig. 11).

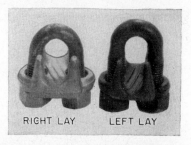

RIGHT LAY   LEFT LAY

Wedge sockets are used on power shovels and similar equipment on which it may be necessary to frequently change the attachment of the rope to the dipper or bucket. The efficiency is low, being only about 70 per cent of the strength of the rope. In using the wedge socket care must be exercised

Fig. 11. A right-lay clip used on a left-lay rope, or vice versa, will cause cutting of the wires in the strands.

to install the rope so that the pulling part is directly in line with the clevis pin (Fig. 12); otherwise a sharp bend will be produced in the rope as it enters the socket.

Socket attachments are used on more permanent installations, such as elevators. In making an attachment the following procedure should be adopted. Put a seizing on the rope at a distance from the end equal to the length of the socket basket, and two additional seizings spaced one and one-half rope diameters apart immediately back of the first seizing. Unlay the strands of the rope, and cut off the hemp center near the first seizing, then carefully unlay and broom out all the wires in the several strands. It is unnecessary to straighten the wires. Clean with carbon tetrachloride or other solvent the wires that have been separated, then dip them for about three-fourths of their length into a 50 per cent solution of commercial muriatic acid for about a half a minute, or until each wire is thoroughly cleaned.   Do not allow the acid to reach the hemp center of the rope.   Then rinse by immersing in boiling water containing a small quantity of bicarbonate of soda to neutralize any remaining trace of the acid.   Use care not to allow acid to come into contact with any other part of the rope.

Bind the wires together, and insert the end of the rope into the socket so that the ends of the wires are even with the top of the basket, and remove this temporary binding wire.   Spread out the wires so as to occupy equally the entire space within the basket.

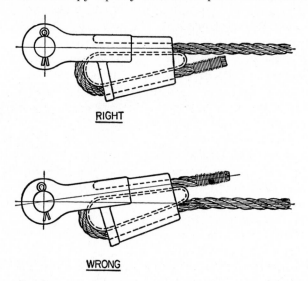

Fig. 12.   The right and wrong ways of placing a wire rope in a wedge socket.

Clay should be applied to the annular space between the neck of the socket and the periphery of the rope to prevent the molten metal from running out.   Hold the socket and several feet of rope in a vertical position, and pour in molten *zinc* until the basket of the socket is full.   The zinc must be at the correct temperature. Before pouring, dip the end of a soft pine stick into the ladle for a few seconds, and remove.   If the metal adheres to the stick, it is too cold.   If it chars the stick, it is too hot.   Allow the socket to cool, then remove the seizings, except the one close to the socket.

Some wire-rope users, elevator manufacturers in particular, prefer to use babbitt in place of zinc.   In preparing the rope for socketing the first seizing is $1\frac{2}{3}$ times the length of the basket from the end of the rope.   Before pouring the metal, bend all the individual wires backward toward the center of the socket so that the bends are about flush with the top of the socket. Do *not* use lead for socketing. Even babbitt sockets have considerably less strength than the zinc sockets, sometimes as low as 25 per cent of the zinc. To check an existing installation scratch the metal with a knife. Zinc is quite hard, babbitt is softer, and lead is very soft.

Babbitt should never be used for hoisting ropes. It is, however, acceptable for use on elevators for the following reasons:

(1) Whereas safety factors of 5 to 7 are commonly used on cranes, etc., the factor of safety on passenger elevators varies from $8\frac{1}{2}$ to $11\frac{3}{4}$, depending upon the speed of the car.

(2) A socket of given metal has the same holding power regardless of the strength of the rope. Hence, if a babbitt socket has an efficiency of, say, 60 per cent of the strength of an improved plow steel rope, it has an efficiency of 87 per cent when used on a traction steel rope of much lower strength.

All hoisting ropes on elevators and mine hoists should be resocketed at frequent intervals, varying from monthly to bi-yearly according to the severity of service. Before resocketing, the socket should be annealed by heating in a wood fire to a cherry-red color and then allowed to cool with the ashes. After the sixth annealing, the socket should be discarded. It should also be noted if the neck or throat of the socket is rounded to reduce the wear and cutting of the rope.

A spliced eye can also be used for attaching the end of a rope. Of course, a heavy-duty thimble is very necessary to avoid bending the rope too sharply around the pin and to reduce the wear on the rope. The efficiency of a well-made eye splice with a heavy-duty thimble varies from about 95 per cent for $\frac{5}{16}$-in. rope, 88 per cent for $\frac{3}{4}$-in. rope, 76 per cent for $1\frac{1}{4}$-in. rope, and 70 per cent for $1\frac{1}{2}$-in. rope. All splices should have at least four tucks, and the completed splice should be carefully wrapped with a wire serving to cover the protruding wire ends to eliminate the danger of lacerations to the hands of those handling the ropes or slings. Tying cable into knots is definitely dangerous.

Methods of splicing wire-rope thimbles vary with the individual habits of mechanics. Some of these methods, however, produce considerable rope distortion at the point of tucking, resulting in a decrease in the ultimate strength of the splice. The following method is a practice in the shops of the John A.

Roebling's Sons Co., and has been made as simple in operation as is consistant with safe and efficient service.

To make a thimble splice in six-strand rope, measure off two to four feet from the end of the rope, depending upon the size of the rope, and bend it at that point. Lay the proper size thimble

Fig. 13.   Step #1 in making an eye splice.   (*John A. Roebling's Sons Co.*)

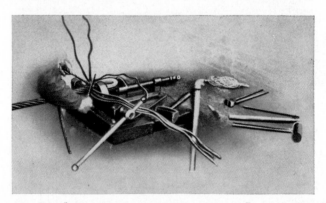

Fig. 14.   Step #2 in making an eye splice.   (*John A. Roebling's Sons Co.*)

into the bend and clamp it in a splicing vise, as illustrated in Fig. 13, with the short end of the rope at the right when looking toward the vise.

The tools are also shown in Fig. 13 and consist of two wooden mallets, a spike, a serving iron, knife, nippers, a piece of pipe, and some fiber rope.

Unlay the rope back to the thimble, serve ends of strands with fine annealed wire, bend four strands to the right and two to the left, and cut out the fiber core as shown in Fig. 14. A strand or independent wire-rope core should be cut in the same manner as a fiber core except not quite so close to the thimble.

Untwist the lay of the rope somewhat by means of the pipe and fiber rope as shown in Fig. 15. Then drive the spike under

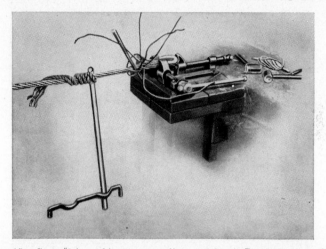

Fig. 15. Step #3 in making an eye splice. (*John A. Roebling's Sons Co.*)

Fig. 16. Identification of individual strands. (*John A. Roebling's Sons Co.*) Fig. 17. Making the first tuck (*John A. Roebling's Sons Co.*)

the two top strands of the main part of the rope and insert the strand lying on top (i.e., under the point of the thimble) through the opening. The spike should be driven from right to left and the strand put through in the same direction. To make this clearer, a section of the rope and of the loose ends is shown in Fig. 16. The strands in the main rope are numbered 1 to 6, and in the loose end *a* to *f* (Fig. 16). The first step then, as stated, is to drive the spike in between 1 and 2, and out between 6 and 5 (Fig. 17) and insert end *a* through this opening. Next pull the

strand up tight and force it back by twisting the spike up along
the lay toward the thimble.

Then tuck the other ends thus: Follow strand 1 around to
where it lies on the bottom of the rope (see Fig. 18) and drive the
spike under it (i.e., between 1 and 6, and 1 and 2), from the
right (Fig. 19) and force end *b* through from the left, bringing

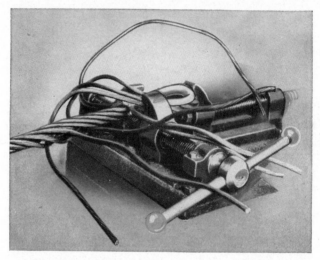

FIG. 18.   Step #4 in making an eye splice.   (*John A. Roebling's Sons Co.*)

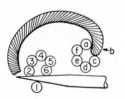 

FIG. 19.   Making the second tuck.   FIG.   20.   Positioning   the   strands.
(*John A. Roebling's Sons Co.*)   (*John A. Roebling's Sons Co.*)

it over the rope.   Twist the spike back along the lay toward the
thimble, after having pulled the end up tight, until the strand
can be worked up no further.   Next drive the spike from the
right under 2 where it is at the bottom of the rope, force end *c*
through from the left and work back as before; and so on, *d*
under 3, *e* under 4, *f* under 5.

All the ends having thus been tucked whole once, they are
tucked whole a second time.   This second tucking consists

merely of wrapping each end once around the same strand under which it was passed in the first tucking operation, and then forcing the tuck back as close to the thimble as possible. Thus *a* lies under 1 and 6 as a result of the first tuck just described (Fig. 17).

Lift 6 by means of spike (Fig. 20), bring *a* around and back under 6 and pull up tight. Strand *b* lies under 1 and is wrapped once around in the same way and so on for *c, d, e,* and *f*. Figure 15 shows the splice while *b* is being tucked the second time. Then strands *a* and *b* are given one more tuck each in exactly the same way as for the second tuck. When this operation is

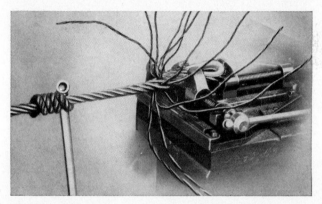

Fɪɢ. 21. Step #5 in making an eye splice. (*John A. Roebling's Sons Co.*)

finished, strands *a* and *b* will have had three whole tucks each and strands *c, d, e,* and *f* two whole tucks each.

All the ends having been tucked whole as described, they are split; that is, by means of a knife the wires are separated into two parallel bundles for each strand, as shown in Fig. 21. One bundle from each end is then wrapped three times around the strand from under which it emerges, in a manner exactly the same as the second tucking of the whole end was done. After this operation is finished, the splice appears as shown in Fig. 22.

Cut off the loose ends as close as possible except the two half strands *a* and *d* as shown in Fig. 22. These two strands should be cut off long enough to reach to the thimble after being opened up and hammered back (Fig. 23).

Take a coil of seizing strand (see Table III), fasten one end loosely to the thimble, and have an assistant hold the coil itself

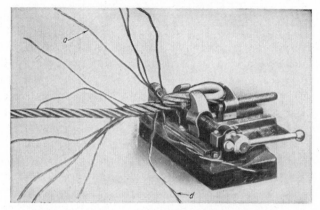

Fig. 22.　Step #6 in making an eye splice.　(*John A. Roebling's Sons Co.*)

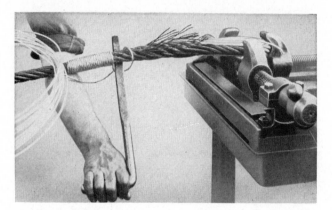

Fig. 23.　Applying the seizing wire.　(*John A. Roebling's Sons Co.*)

TABLE III.　SEIZING STRAND FOR THIMBLE SPLICES

(7-wire galvanized seizing strand.)

Rope Diameter, in.	*Strand Diameter, in.*
$\frac{1}{4}$	$\frac{1}{16}$
$\frac{5}{16}$	$\frac{1}{16}$
$\frac{3}{8}$	$\frac{1}{16}$
$\frac{1}{2}$	$\frac{3}{32}$
$\frac{5}{8}$	$\frac{1}{8}$
$\frac{3}{4}$	$\frac{1}{8}$
$\frac{7}{8}$	$\frac{5}{32}$
1	$\frac{5}{32}$
$1\frac{1}{8}$	$\frac{5}{32}$
$1\frac{1}{4}$ and larger	$\frac{3}{16}$

in the location shown in Fig. 23. Then, using the serving iron as illustrated in the same illustration, apply the serving. When up to the thimble, twist the ends together and cut off.

When completed the thimble splice will appear as in Fig. 24. If properly made and correctly used, there is no danger of it ever pulling out; and under excessive load such a splice would develop about 60 to 90 per cent of the ultimate strength of the rope, depending upon the size. The only reason it will not carry up to the full breaking strength is that the wires nick each other under heavy stress where the strands cross inside the tuck and are weakened slightly. The weakest part of the splice is in the vicinity of the last set of tucks, and for this reason it is very important not to hammer, or otherwise distort this section, more than is absolutely necessary.

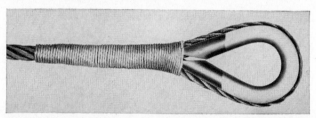

FIG. 24.   The finished eye splice.   (*John A. Roebling's Sons Co.*)

If no seizing strand is to be used, all strands should have four instead of three tucks. All ends are then cut short and the finished splice is hammered lightly.

Mechanical eye splices (Fig. 25) will develop, their manufacturers claim, 100 per cent of the strength of the rope, as indicated in the catalogue table of strengths. Do not be confused and think that the splice is stronger than the rope. The fact is that the rope is always stronger than indicated in the table. Therefore it is possible for the splice to be as strong as the catalogue strength of the rope, and still fail before the body of the rope fails.

One manufacturer (Fig. 25a) bends the end of the rope into a J shape, then unlays or separates the six strands and relays them into the grooves between the strands around the body of the rope. To hold these strands in place two steel ferrules are pressed onto the splice by a hydraulic press.

Other manufacturers (Fig. 25b) bend the rope into the J shape, with the end placed against the body of the rope. An aluminum ferrule is pressed on to hold the parts together, the aluminum being cold-flowed around the strands and wires to make one solid mass. Steel ferrules used on this type of splice have been known to pull out while in service. Hence, the author is reluctant to recommend this type of mechanical splice having a steel ferrule. Tests have proven that when an aluminum ferrule is used this type of splice will fail at a load greater than the catalogue strength of the rope.

Still other manufacturers first make a Flemish eye (Fig. 25c), as described below. But instead of using friction tape to secure

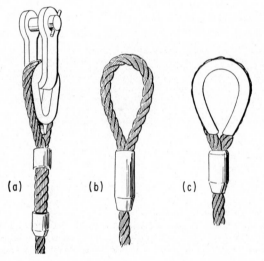

(a)　　　　(b)　　　　(c)

Fig. 25. Mechanical eye splices of several makes for wire rope: (a) American Chain & Cable Co. "Duraloc." (b) Jones and Laughlin Steel Co. "Jal-Klamp." (c) John A. Roebling's Sons Co. Flemish eye.

the strand ends to the body of the rope, the ends of the six strands are separated and each laid in one of the helical grooves between the strands in the rope. Then, with a force of about 500 tons, a steel ferrule is squeezed onto the splice. As indicated below, a Flemish eye splice has great strength even without the ferrule. The ferrule, however, gives the additional holding power for the strands to make the breaking strength of the splice as high as the catalogue strength of the rope itself. Of the various types of mechanical splices this is undoubtedly the most dependable. These mechanical splices can only be made in the plant of a wire rope or sling manufacturer.

Occasionally it becomes necessary to do a special job, such as snaking a small-size wire rope through an underground pipe or conduit to pull into place an electric cable. For attachment to the wire an eye is required, but clips cannot be used because they take up too much room and they might catch on joints in the pipe. A spliced eye is desirable, but perhaps there is available no experienced splicer, nor the splicer's tools needed to make a conventional splice.

For such a job a Flemish eye may be used. This can be easily and quickly made by a man with no previous splicing experience.

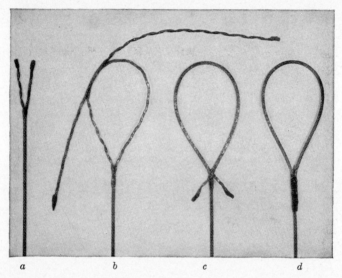

a   *b*   *c*   *d*

Fig. 26. The Flemish eye splice can be made without tools and with little experience.

The Flemish eye develops only about 70 per cent of the strength of the rope, and consequently should not be used for hoisting loads.

Using a *preformed* rope of the required size, make four wraps of friction tape around it at a point about 20 in. from the end. Then unlay together three adjacent strands, leaving the other three strands and the hemp or wire rope center. Bind their respective ends together with four wraps of friction tape, using care to keep the strands in their correct positions relative to each other (Fig. 26a).

Just short of midway between the crotch and the ends of the

strands, tie the groups of strands into what appears to be the first step in making a square knot, and you will find that the strands fit right into the grooves from which they were removed, but running in the opposite direction (Fig. 26*b*). Continue laying the strands into the groove until the crotch is reached and the last possible wrap completed (Fig. 26*c*). Then remove the tape seizing from the strand ends, separate the strands, and lay them orderly around the rope in the grooves between the strands; apply a seizing of friction tape to hold these wires in place (Fig. 26*d*) and the job is done.

Standard wire guys are frequently used by electric utility companies to support their wooden line poles against side pull. In the past, to attach a guy to the eyebolt at the pole top, or to the ground anchor, or to a porcelain strain insulator in mid-air, it has been the practice to make an eye by using a 3-bolt plate

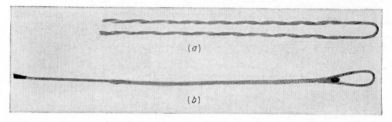

(*a*)

(*b*)

Fig. 26A. Preformed grip for use in attaching wire guy to eyebolt of ground anchor. (*Preformed Line Products Co.*)

clamp.   There is now available a preformed wire guy grip which can be installed quickly and easily without the use of special tools.

To attach the lower end of a guy to, let us say, the eyebolt attached to a ground anchor a "come-along" or other tension-producing device is secured to the guy and to the anchor rod to pull the guy taut.   If so required, insert a thimble into the eyebolt.   Make several wraps of friction tape around the guy at a point about 4 in. from the eyebolt, and cut off the guy just below this tape seizing.

Insert one leg of the guy grip (Fig. 26A-*a*) into the eyebolt (and bearing on the thimble, if used).   Rotate one leg of the grip counterclockwise around the guy, while pulling outward away from the guy with enough force to permit the leg to form a spiral around the guy.   Wrap the leg around the guy three turns only, then start wrapping the other leg around the guy so as to lie in the open space between the wraps of the first leg.   Then finish the job by simultaneously laying the remainder of the two clamp legs, one in each hand.   The finished job is shown in Fig. 26A-*b*. It is claimed that the holding power of the grip is about equal to the catalog strength of the guy.

Sheaves.   Undersize sheaves are probably directly responsible for more rope failures than any other single cause.   Under no condition should a rope be operated over a sheave smaller than the "critical" diameter, as sharper bends result in displacement of the strands and overstressing of the wires.   The minimum safe

TABLE IV.   SHEAVE DIAMETER*

Rope construction	Minimum diameter†	Critical diameter†
6 × 19 Seale	34*d*	20*d*
6 × 16 filler wire	30*d*	16*d*
6 × 19 Warrington	30*d*	16*d*
Flattened strand	30*d*	
8 × 19 Seale	26*d*	16*d*
8 × 19 filler wire	26*d*	
6 × 22 filler wire	23*d*	
8 × 19 Warrington	21*d*	14*d*
8 × 19 filler wire	21*d*	
6 × 37 Seale	18*d*	14*d*
6 × 41	18*d*	
6 × 6 × 7 tiller rope	. . .	10*d*

* Courtesy of American Chain and Cable Co.
† The minimum diameter of sheave for satisfactory operation, and the critical diameter where rope is damaged.   *d* is the rope diameter.

operating and the critical diameters given by rope manufacturers are indicated in Table IV, where *d* is the nominal rope diameter.

Figure 27 presents data for comparing the service life of a wire rope which you are now using with one which you propose to use. For instance, the records show that on a certain crane the $\frac{1}{2}$-in. 6 × 19 Seale lay rope has a useful life of about 5 years.   In the table we find that this rope has an index of 0.80.   The rope proposed is a $\frac{1}{2}$-in. 6 × 25 filler-wire type, having an index of 1.00. Under the same operating conditions the proposed rope may be expected to have a useful life of 5 years $\times \dfrac{1.00}{0.80} = 6\frac{1}{4}$ years.

However, the fact that a rope with a higher index has smaller outside wires to resist wear and abrasion may be a factor.

Likewise, if small sheaves are suspected of causing the short life we may check with the curve in Fig. 27. Say that the

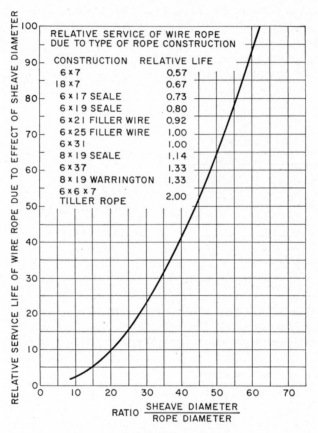

FIG. 27. Relative life of wire rope.

present $\frac{1}{2}$-in. rope operates over sheaves of 8-in.-groove diameter. Then the ratio of sheave diameter to rope diameter is $\frac{8''}{\frac{1}{2}''} = 16$, and from the curve we find the index to be $6\frac{1}{2}$. Operating conditions will permit the installation of 12-in. sheaves, so we find that the ratio is $\frac{12''}{\frac{1}{2}''} = 24$, and the index is about 14. In other words, the use of the larger sheaves may be expected to increase the rope life by $14:6\frac{1}{2}$, and the rope should last twice as long.

The sheave groove must be large enough to accommodate a new rope (which usually is slightly oversize) and should be within the tolerance indicated in Table V.

Corrugated grooves will cause a rapid depreciation of a rope. Also, a badly worn sheave groove, which is caused by an old rope, will pinch and wear a new rope when it is installed. To accommodate a new rope the groove should be larger than the nominal diameter of the rope. Test gauges made to the diameters indicated in Table V can be used to check the size of the grooves.

TABLE V. SHEAVE GROOVES*

Rope diameter, in.	Tolerances, in.
$\frac{1}{4}$ to $\frac{1}{2}$	$+\frac{1}{32}$ to $+\frac{3}{32}$
$\frac{9}{16}$ to 1	$+\frac{1}{16}$ to $+\frac{1}{8}$
$1\frac{1}{8}$ to 2	$+\frac{3}{32}$ to $+\frac{3}{16}$

* Proper groove diameter for sheave is very important.

In purchasing sheaves, the nominal diameter frequently indicates the over-all flange diameter, but in this book the sheave diameter referred to is the diameter at the bottom of the groove, unless otherwise indicated.

Reverse bends in a rope should be avoided so far as possible, but where they are unavoidable, the sheaves causing the reverse bend should be spaced as far apart as practicable (*A* and *B*, Fig. 28) and the most offending sheave should be made about one-third to one-half larger than the others.

Even the equalizing sheave on a hoist block should be kept larger than the critical diameter, for although this is not an operating sheave in the ordinary sense of the word, there is a certain amount of motion here that may cause fatigue failure at this point even before the rope fails elsewhere.

In the design of sheaves consideration must also be given to the bearing pressure of the rope on the sheave groove. If the rope bears too heavily on the groove of the sheave, and if the metal is softer than the wire rope, the former will become corrugated to fit the contour of the rope strands. These corrugations, in turn, will help to wear the rope, particularly if a new rope is run over a sheave having the groove corrugated to fit a worn and undersize rope.

The following formula is used to determine the rope bearing pressure:

$$P = \frac{2t}{D \times d}$$

where P = bearing pressure, lb per sq in.
t = rope tension, lb
D = tread diameter of sheave, in.
d = rope diameter, in.

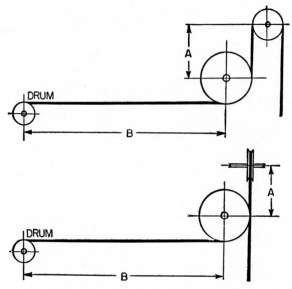

Fig. 28.   Two types of rope bends that shorten the life of a wire rope.

TABLE VI.   ALLOWABLE BEARING PRESSURES*

	Lb per sq in.
Cast-iron sheaves.....................	600
Steel casting sheaves.................	1,075
Chilled cast-iron sheaves.............	1,325
Manganese steel sheaves.............	3,000

* The pressure of the rope on the sheave groove should not exceed the allowable values.

For regular-lay ropes the bearing pressure should not materially exceed the values given in Table VI.

Drums.   Drums should be designed to conform to the rules for sheaves, although a drum can be made slightly smaller than the minimum recommended size for sheaves owing to the fact that a rope is flexed only once by the drum whereas it is flexed twice, bending and straightening, each time it passes over the sheave.

As far as is practicable, a drum should be designed to accommodate all the rope on one smooth, even layer. Two and sometimes three layers are permitted, but more than three layers may cause crushing of the rope on the bottom layer as well as at the end of any layer where pinching occurs.

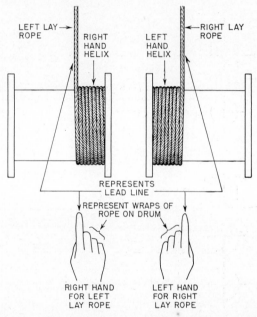

FIG. 29.   The proper way to wind wire rope on a drum, using the hand as a guide.

There is a proper way to wind rope on a drum, and this method should be followed so far as practicable.   If right-lay rope is used, as is generally the case, it should wind on the drum in a left-hand helix; if left-lay rope is used, it should wind in a right-hand helix. Figure 29 shows a rule-of-thumb, or should we say a "rule-of-hand," which can be readily memorized.   Stand behind the drum, looking toward the lead sheave.   In most installations the rope winds onto the top of the drum.   If right-lay rope is used hold the left-hand palm downward with the fingers and thumb clinched, except the index finger which points in the direction of the lead line.   The index finger represents the lead line; the bent fingers represent the wraps of rope on the drum.   For left-lay rope use the right hand.   If the rope leads from the bottom of the drum hold the hand with the palm upward.

Fleet Angle.   In a properly designed installation the point where the cable leaves one sheave must lie in the plane of the sheave toward which the rope is leading.   As in all cases except rope drives the rope runs in both directions, this rule applies to all sheaves.   The lead sheaves likewise should line up with the center

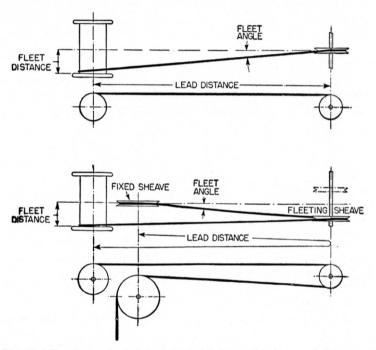

Fig. 30.   The fleet or lead distance is that between the drum and the nearest fixed sheave, measured along the path of the rope.

of the width of the drum.   The variance from this line is called the "fleet angle" and in the case of a smooth drum should not exceed $1\frac{1}{2}$ deg, or 1 in. in 40 in.   Two degrees, or $1:28$, is permissible with a grooved or scored drum (see Fig. 30).

If the sheave nearest to the drum can fleet or slide along a fixed shaft, the lead distance is the length of the rope between the drum and the nearest fixed sheave.

Bending Stress.   Although there is some difference of opinion relative to the exact stresses produced by bending a rope over a sheave, the fact remains that stresses of considerable magnitude are developed, and in some cases the bending stress may exceed that caused by the live load. The handbooks of the various rope manufacturers give rules for calculating the bending stress. The factor of safety must take care of these unknown stresses, assuming that the sheaves are equal to or larger than the recommended minimum size.

Displacement of Strands. Any displacement of strands from their original positions will cause a noticeable loss of strength of the rope, because under this condition some strands will take more than their share of the load. If the rope has been crushed or kinked or bent too sharply, it will be flattened and the load will not be equally distributed on the strands.

If a rope has been improperly seized one of the strands may slip and loosen, thus permitting this strand to be forced out of its positions by the adjacent strands and giving the rope a corkscrew appearance. A Lang-lay rope that has not been properly seized may open up and produce a condition known as "bird-caging."

The pitch of a used rope should not differ materially from that of a new rope, which is generally about $6\frac{3}{4}$ times the rope diameter. In some cases, owing to improper sheaves, that portion of the rope which continually passes over them may have its lay increased, the twists being accumulated at the ends of the rope, as evidenced by the shortening of the lay at these points. Lengthening of the lay at a socket is an indication of an improper socketing job; the end of such a rope should be cut off, and a new socketing job made.

Sudden Stresses. Consideration must be given to the stress produced by rapid acceleration and deceleration, jerks, vibration, etc. An automotive crane traveling over a rough road with a load suspended on the boom may have impact stresses of 100 per cent or more added to the sum of the live and dead loads.

If a load is accelerated rapidly, such as in the case of a high-lift coal-unloading tower, the acceleration stress can reach great magnitude. Let us assume the following conditions:

Weight of coal and bucket, 10,500 lb.

Hoisting speed, 1,800 ft per min = 30 ft per sec.

Time to accelerate from rest to full speed, $1\frac{1}{2}$ sec.

While at rest, the load on the hoist cable is 10,500 lb. To this must be added the force necessary to accelerate the load. This is obtained from the formula

$$f = \frac{wv}{32.16t}$$

where f = accelerating force, lb

w = weight to be lifted, or static tension, on rope, lb

v = velocity, ft per sec

t = time to reach full speed, sec

$$f = \frac{10,500 \times 30}{32.16 \times 1\frac{1}{2}} = 6,530 \text{ lb in this example}$$

Thus, to the static tension of 10,500 lb on the rope must be added the acceleration stress of 6,540 lb, making a total of 17,030 lb.

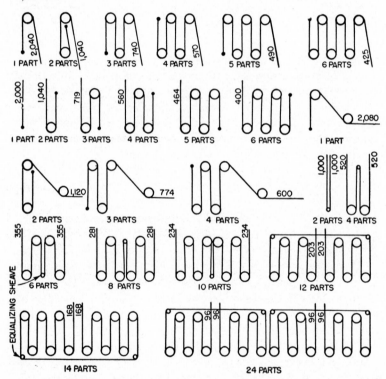

FIG. 31. The pull on the lead line when hoisting a load of 1 ton, and when using various reevings.

Friction Load. The static tension on a rope is the total weight of the load, load block, hook, slings, etc., divided by the number of parts of rope actually supporting the load block.

As a sheave is rotated on its shaft, friction must be overcome. Also, friction is developed when the wires and strands of a rope slide on each other as the rope is flexed in passing on to and off a sheave.

Figure 31 gives the approximate tension on the rope for various rigging methods.   The figures given represent the pull in pounds to lift a gross load of 1 ton, exclusive of acceleration stress.

Total Stress.   The total stress imposed on the operating portion of a wire rope may consist of many or all of the following items:

1. Live load to be hoisted
2. Dead load (lower load block, hook, slings, cable)
3. Additional stress required to overcome friction of sheaves
4. Sudden load due to quick acceleration, deceleration, or shocks
5. Bending stress due to bending over sheaves and drum

Friction on Drum.   In order to reduce to a safe figure the load on the cable anchorage on the drum, it is necessary to leave at

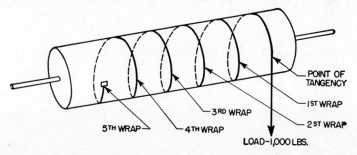

Fɪɢ. 32.   The strain on the anchorage of the cable on the drum depends upon the number of wraps of the rope on the drum and upon the lubrication.

least two and preferably three complete wraps of rope on the drum when the load block is at the floor level or lowest position.   Figure 32 and Table VII indicate the strain at the anchorage in pounds for each 1,000-lb tension in the rope, depending upon the number of wraps on the drum and upon the lubricant on the rope.

Breaking Strength of Wire Rope.   Table VIII gives the breaking strength of 6 × 19 wire ropes of various grades of steel.

For ropes of 6 × 7 type use $96\frac{1}{2}$ per cent of the tabulated strengths.

For ropes of 6 × 37 class use 95 per cent of the tabulated strengths.

For ropes of 18 × 7 type use $91\frac{1}{2}$ per cent of the tabulated strengths.

For ropes with independent wire-rope center use 107 per cent of the tabulated strengths and 110 per cent of the tabulated weights.

TABLE VII. STRAIN ON ROPE AT GIVEN NUMBER OF WRAPS FROM
POINT OF TANGENCY

(1,000-lb load)

Number of wraps	Well-greased rope, lb $f = 0.07$	Dry rope, lb $f = 0.17$
$\frac{1}{4}$	890	760
$\frac{1}{2}$	788	575
$\frac{3}{4}$	697	361
1	616	329
$1\frac{1}{2}$	485	188
2	369	101
$2\frac{1}{2}$	287	59.7
3	220	33.9
$3\frac{1}{2}$	166	18.4
4	126	10.1
$4\frac{1}{2}$	95.3	5.8
5	71.6	3.1

TABLE VIII. BREAKING STRENGTH OF WIRE ROPE

(Divide by factor of safety in Table IX to obtain safe load)

Diameter, in.	Mild plow steel, lb	Plow steel, lb	Improved, plow steel, lb	Extra improved plow steel, lb	Weight per foot, lb
$\frac{1}{4}$	4,140	4,780	5,480	6,040	0.10
$\frac{5}{16}$	6,440	7,420	8,520	9,380	0.16
$\frac{3}{8}$	9,240	10,620	12,200	13,420	0.23
$\frac{7}{16}$	12,500	14,380	16,540	18,180	0.31
$\frac{1}{2}$	16,260	18,700	21,400	23,600	0.40
$\frac{9}{16}$	20,400	23,600	27,000	29,800	0.51
$\frac{5}{8}$	25,200	29,000	33,400	36,600	0.63
$\frac{3}{4}$	36,000	41,400	47,600	52,400	0.90
$\frac{7}{8}$	48,600	56,000	64,400	70,800	1.20
1	63,200	72,800	83,600	92,000	1.60
$1\frac{1}{8}$	79,600	91,400	105,200	115,800	2.00
$1\frac{1}{4}$	97,600	112,400	129,200	142,000	2.50

Ropes having an independent wire-rope center are about $7\frac{1}{2}$ to 10 per cent stronger than indicated above.

Factors of Safety. In order to ensure against failure of a wire rope in service with possible disastrous consequences, the actual strain on the rope should be only a fraction of the breaking load.

The breaking strength divided by the actual total stress on the rope is known as the "factor of safety" and should not be less than the values given in Table IX.

TABLE IX. MINIMUM SAFETY FACTORS FOR WIRE ROPE

Slings	8
Overhead electric hoists (small)	7
Overhead traveling cranes (small)	$6\frac{1}{2}$
Industrial truck cranes	6
Locomotive cranes	6
Derricks	6
Grab buckets	6
Overhead traveling cranes (large)	$5\frac{1}{2}$
Hoisting tackle	5
Shovels	5
Derrick guys	5
Stack guys	4
Tow ropes	4
Ferry ropes	4
Cableway track ropes	4
Contractor's suspension bridges	3
Elevators	See local ordinances

Breaking In a New Rope. After installing a new rope, operate it for perhaps an hour without the "live" load to ensure that it will accommodate itself to the sheaves and drums before the heavy strain is applied. The time thus spent will ensure longer operation without servicing. As mentioned above, the clips should be tightened after the breaking-in period.

After about half of the normal life of the rope, remove it from the equipment, turn it end for end, and reinstall it. The useful life will be much longer than if the rope is left in its original position. Or have a rope long enough to periodically cut off a 20- to 25-ft length.

On large coal-hoisting apparatus and on other equipment where the life of a wire rope is relatively short, the frequent removal of the worn rope, and the installation of a new rope by reeving it through the many fixed sheaves and through the sheaves in the grab bucket, is time- and labor-consuming. As an acetylene welding outfit is usually available at every rigging job, it has been found practicable to weld the end of the new rope to the end of the worn rope (Fig. 33), then by winding the old rope onto its drum the new rope is quickly reeved through all the sheaves. Do not, under any condition, apply a live load to a rope thus welded.

Although this attachment does not have the strength of the rope, it does have more than ample strength to pull the new rope into place. But a word of warning: before starting to weld the ropes together, the strands of each rope must be unlaid for 6 in. or more and the ends of the hemp centers cut off and removed. Also, in order to prevent the flame of the torch from flashing along the entire length of the grease-coated ropes, a few wraps of

Fig. 33. Welding the new rope to the worn rope facilitates reeving it through the many sheaves.

asbestos tape should be placed a short distance each side of the weld.

Special Rigging. In hoisting operations an improvised support is frequently made by stretching a wire rope more or less tightly between two beams, roof trusses, or other elevated structural members and attaching a chain hoist at some point in its length. This method of suspending a chain hoist or other load cannot be too strongly condemned, as in practically every instance the loads produced cause excessive stresses both in the cable and in the structural supports. This is particularly objectionable in the case of the structural members, as they are subjected to forces in a direction for which they probably were not designed.

In order that the rigger may approximate the loads imposed on the cable (and on the structural members) by such a rigging, the following instructions should be followed.

Draw to any suitable scale an elevation of the structural supports and the cable stretched between them, with particular attention to the amount of sag of the cable, as shown in Fig. 34.

Next construct a triangle of forces by first drawing a vertical line *DE* to represent the vertical load, which includes the weight of the load to be lifted and the weight of the chain block.   Draw this vertical line to any convenient scale, such as $\frac{1}{2}$ in. = 1,000 lb, 1 in. = 1,000 lb, 1 in. = 10,000 lb, etc.   Then draw a line *DF* parallel to the portion of the cable *AB*, starting at *D* and continuing indefinitely.   Next draw a line *EG* parallel to the portion of the cable *BC*, starting at *E* and continuing until it crosses line *DF* at *H*.

Now by measuring the lines *DH* and *EH* with the same scale used for line *DE*, the stresses can be determined.   In this example, the stresses are 14,200 lb in the portion of the cable *AB* and

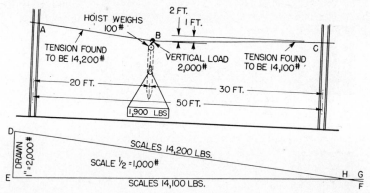

Fig. 34.   Triangle of forces is used to calculate the stress in a tighly stretched trolley cable.

14,100 in *BC*.   These forces are imposed nearly horizontally on the structural members.

Thus, it will be seen that a comparatively light load of 2,000 lb suspended by the cable will impose a horizontal force of over 14,000 lb on the beams, columns, or other structural supports. If there are any shocks or impact stresses, these will be likewise amplified and transmitted to the steelwork.

Likewise, when slings are used as spreaders for picking up loads, their angles are of great importance, as shown in the following sketch.   To pick up a load of 2,000 lb on two ropes that are parallel, each rope will be stressed to 1,000 lb (Fig. 35).   If these slings are attached to a common hook on the hoist so that their angle to the horizontal is reduced to 60 deg, the stress is increased to 1,155 lb.   At 45 deg the stress is 1,414 lb, at 30 deg it is 2,000 lb, and at 5 deg the stress has reached 11,470 lb.   Not only is there danger of overloading the sling cable and causing its failure, but the crate, box, or even the load itself may be crushed by the force applied at its upper corners.   In any event, padding should be provided to protect the slings from sharp bending and possible cutting at all corners of the load.

Inspection.   One of the difficult problems confronting the equipment inspector is deciding just when a wire rope has reached the limit of its safe usage and must be discarded.   Naturally, it is poor economy to discard an expensive hoisting rope before it is necessary.   Likewise, it is dangerous (and it may also prove even more expensive) to continue its use beyond a certain stage.

To determine the proper time to condemn the rope a careful inspection should be made not only of the rope itself but also of the sheaves and other parts that affect its use.

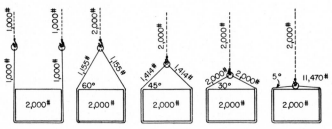

FIG. 35.   The tension on a sling rope (or chain) depends upon its angle as well as upon the load to be lifted.

Some of these factors are as follows:

1. *Construction of the Rope.*   Is the rope of the proper type?

2. *Broken Wires.*   The total number of broken wires in all the strands within a distance of one rope lay (the distance in which a strand makes one complete turn around the rope) at the worst portion of the rope is taken as an index of the reduction in strength due to this cause.   In inspecting preformed rope extra care should be exercised, as the broken wires lie flat in position and are often difficult to detect.   See Tables X and XI for the number of broken wires allowed for various types of rope with varying amounts of wear.

3. *Location of Wire Breaks.*   It should be observed whether or not several adjacent wires are broken, whether the breaks occur at the point of tangency with the sheave groove (which may indicate fatigue of the metal) or with the adjacent strands (which

usually indicate lack of internal lubrication), whether the breaks are about equally distributed among the strands or are mostly located in a few strands. Unequal distribution of the breaks results in a greater loss of strength than when the same number of breaks are equally distributed (see Tables X and XI). Even a single broken wire at the throat of a socket is sufficient cause for cutting off the end of the rope and making a new socket attachment.

TABLE X.   ALLOWABLE WIRE BREAKS AND WEAR ON MINE HOIST ROPES*

Broken Wires in One Rope Lay		Wear on Outside Wires, %
None	but	35
3	or	30
4	or	20
5	or	10
6	and	0

* U.S. Bureau of Mines.

4. *Wear on Outside Wires.*   The length of the shiny, worn spots (Fig. 36) on the outside wires should be measured with a steel scale graduated in one-hundredths of an inch, as this is an index of the reduction in area, and likewise in strength of the wires and of the rope due to wear.

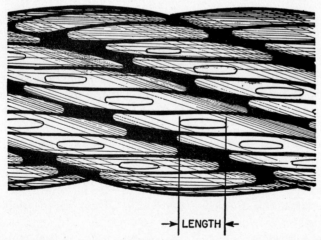

LENGTH

FIG. 36.   The length of the long shiny spots on the wires is an indication of the amount of wear.

Where the personal and property hazard (see Item 24) is great, such as in elevators, mine hoists, and hot-metal cranes, a very strict rule should be followed for condemning a rope, such as the rule of the U.S. Bureau of Mines, which is given in Table X.

TABLE XI.   ALLOWABLE WIRE BREAKS AND WEAR ON
HOISTING ROPES*

Use	Conditions		6 × 19 Warr.	6 × 37	8 × 19 Warr.	6 × 19 Seale	Other conditions
Cranes....	{	Total	18	30	24	......	} Wear on outside wires in excess of 33%
	{	Adjacent	4	6	4	......	
Hoists....	{	Total	24	......	32	......	
	{	Adjacent	5	......	5	......	
Passenger elevators	{	Total	12	......	16	8	} Wear on outside wires in excess of 33%
	{	Adjacent	4	......	4	3	
Freight elevators	{	Total	13	......	18	9	} Wear on outside wires in excess of 33%
	{	Adjacent	4	......	4	3	

* American Cable Co. recommendations for condemning worn wire rope.

On the other hand, where personal and property hazard is not quite so great, such as in the case of bucket hoists, skip hoists, unloaders, cranes, and derricks, a more liberal rule can be followed in condemning wire rope, such as is given in Table XI. The term "total" indicates the total number of wire breaks in all strands in one rope lay. "Adjacent" indicates the number of adjacent wires broken.

On machinery such as horizontal cable roads, drag lines, and steam shovels, where no danger will ordinarily result from a cable failure, it is sometimes customary to keep the rope in service until an entire strand is about ready to fail.

5. *Reserve Strength.* This factor is the ratio of the cross-sectional area of all the inside wires to the area of the rope. In other words, in case of failure of all the outside wires, the strength of the inner wires should remain as a reserve. Seale-lay ropes have the lowest reserve strength, and 37-wire-strand ropes the highest.

6. *Corrosion*

7. *Kinks*

8. *Crushed Strands*
9. *High Stranding*
10. *Bird-caging*
11. *Pitch of Rope Lay*
12. *Attachments*
13. *Splices*

14. *Turns in the Rope.* In the case of hoists, derricks, etc., having multiple reeving of the rope in the pulley blocks, the rope during its early life will stretch and unlay slightly, causing the load block to rotate and the several parts of the ropes to twist around themselves. In such a case the end of the hoist rope, whether it be at the load block or at the boom head, should be detached and rotated so as to remove the turns. In the case of a new rope a few more turns than are necessary can be made in the rope in anticipation of the subsequent stretch.

15. *Frequency of Use.* The number of hours per day or week during which the rope is in use is an important factor. The inspector should consider how much use and depreciation a rope may be expected to have before the next inspection. In other words, consider two ropes in identical conditions and under similar loads. One rope is used only an hour or so a week and probably will be safe for use at least until the next inspection, while the other rope, which is in use 8 hr a day, will probably be in a dangerous condition in a much shorter time. Therefore, the much-used rope might be condemned while the other might be accepted for further use.

16. *Rope Speed.* A high-speed rope will have a much shorter life and also expectancy of life than a slow-speed rope and therefore should be given a more severe inspection.

17. *Rope Bends*
18. *Sheave and Drum Grooves*
19. *Sudden Loads and Shocks*
20. *Equalizing Sheave*
21. *Multiple Winding on Drum*
22. *Lubrication*

23. *Weak Points.* The weakest part of a rope, like the proverbial chain, determines the strength of the rope. This weak point may be where the rope is badly worn, where it has a number of broken wires, where the stresses are unequal in the strands, where the local stresses due to bending the rope over a sheave are excessive, or where the rope is attached to the equipment upon which

it is used.   For instance, if a $\frac{3}{4}$-in. plow steel rope has a safe load of 9,200 lb under direct tension, and if the bending stress at the sheaves reduces the strength to, say, 75 per cent of this load, then there is no need to be concerned over the use of a clipped eye that has an efficiency of about 80 per cent, for it is still stronger than that portion of the rope passing over the sheave.   Also, the strength of the rope itself will eventually be reduced by wear, while the strength of the attachment should remain unchanged.

24. *Personal and Property Hazard.*   Consideration should be given to the possible consequences resulting from failure of the rope, especially where men are hoisted, as in elevators or mine cages, or where heavy loads are carried above men or valuable machinery.

25. *Original Safety Factor.*   Where a liberal factor of safety had been allowed when the rope was new, naturally a greater depreciation can be allowed than if a low initial safety factor had been used.   Each of the above factors should be thoroughly investigated.   This means inspecting every foot of the length of the cable, as the spot that is overlooked may contain the weakest part of the rope.   These factors are then summed up in relation to the personal and property hazard and the original safety factor in order to determine whether the rope should be continued in service or condemned.

Where ropes habitually depreciate rapidly at one end owing to the character of the equipment upon which they operate, it is economical to reverse the ropes end for end before they become bad enough to condemn.

It is recommended that all wire rope issued from the company storehouse be identified by a metal tag bearing a serial number that will, in turn, afford such pertinent information as the date issued, quality of the rope, manufacturer, etc.   Some organizations use a 1-in. length of aluminum tubing, just large enough to slip over one strand of the rope at the seizing, socket, or eye splice, with the serial number stamped on it.   Where it is not desirable to do this, only one grade of rope should be kept in stock, for if several grades are stocked, there will be no means of differentiating between them once they are removed from the manufacturer's spools, and accidents may result from assuming a rope to be of a higher grade than it actually is.

As can be learned from the above text, properly inspecting a wire rope is a man-size job.   It is only after years of experience, together with a good knowledge of the technical properties of rope, that an inspector can learn just when it is economical to condemn a wire rope and still be thoroughly safety-minded.

Investigating Rope Failures. The manner in which the individual wires of a wire rope break should give an indication of the probable cause of failure. For instance, if the wires are badly flattened it is evident that wear or abrasion was a factor. If the breaks are square across, the probable cause is fatigue produced by operation over undersize sheaves and/or numerous reverse bends. Failure due to overloading is usually indicated by the wires necking down at the point of failure.

CHAPTER IV

HOISTING CHAINS, HOOKS, ETC.

The manufacture and use of chain dates back to centuries before the Christian era. Modern chains are the result of many improvements in the material and in the method of manufacture, alloy steel chains having the greatest strength. Originally chains were used almost exclusively for heavy hoisting, such as on cranes, but today they have been largely superseded by wire rope.

There are jobs, however, on which chain is better suited than wire rope, and consequently many chains are in service today. They are particularly well suited for slings for lifting rough loads such as heavy castings, the handling of which would quickly destroy wire-rope slings, owing to bending them sharply over the edges of the castings. Chains are also used extensively for dredging and other marine work, as they will withstand abrasion and corrosion better than cable.

A link of a chain consists of two sides, either of which in failing would cause the link to open and drop the load. A wire rope, on the other hand, is frequently composed of 114 individual wires, all of which must fail before it parts. Chain may thus be said to have less reserve strength, and it should therefore be more carefully inspected. In the case of manually welded chain, the welds depend largely upon the human element.

When a wire rope is fatigued from severe service, the wires break one after the other over a relatively long period of time and thus afford the inspector an opportunity to discover the condition. If severely overloaded, the wires and strands will break progressively over a period of perhaps several seconds, and with resultant noise, before complete failure occurs. This may afford the man handling the load a brief time in which to jump to safety before it is dropped.

Chains, on the other hand, will usually stretch under excessive loading, the links elongating and narrowing down until they bind on each other, thus giving a visible warning. However, if the overloading is great enough, the chain will ultimately fail with less warning than wire rope. Should the weld be defective, a chain may break with little, if any, warning.

84

Owing to the construction of wire rope, with the individual wires running in a helical direction, it has more stretch and will therefore be more resistant to sudden loads or shocks than will chain.

According to the Chain Institute:

The safe working load is the maximum load in pounds which at any time or under any condition should be applied to the chain or the attachment thereon, even when the chain is in the same condition it was in when it left the factory, and even when the load is applied only in direct tension to a straight length of chain or to an attachment thereon. . . . Any change in the above factors, such as twisting of the chain, deterioration of the chain or attachment thereon by strain, by usage, by weathering, or by lapse of time, or acceleration in the rate of application of the load or variation in the angle of the load to some sharper angle resulting from the configuration or structure of the material constituting the load will lessen the load that the chain or the attachment thereon will safely withstand.

Table I gives the working loads for various types of chain, based on a factor of safety of about $3\frac{1}{3}$ to 4, that are recommended by many chain manufacturers. Yale & Towne Mfg. Co., however, uses a safety factor of $5\frac{1}{2}$ to $6\frac{1}{2}$ on their $\frac{1}{2}$- to 2-ton chain hoists. Considering the fact that a minimum factor of safety of 5 is recommended for wire rope, and considering the limitations indicated in the previous paragraph, the author advises working loads, particularly for slings, not much in excess of one-half the values given in the table (except Yale & Towne chain).

The tabular working loads are undoubtedly safe for straight pull, as stated by the Chain Institute, but there is no assurance that the weight of the load is not underestimated, that the chain has not been weakened by usage or by weathering, or that there will not be accidental shock or impact loading. To cover all these possibilities the higher safety factor is recommended.

Iron hoisting chains should be annealed every two years to relieve work hardening; those used as slings should be annealed annually. After six annealings, the chains should no longer be used for hoisting purposes. Steel chains should not be heat-treated after leaving the factory.

Inspection. Let us consider the inspection of chain from a safety viewpoint. All chains should be thoroughly inspected at least once a month by a competent inspector. Every chain, whether attached to a piece of equipment or used as a sling, should be given a serial number, which can be stamped on a metal tag attached to the end link or ring. A book record should be kept

TABLE I. SAFE WORKING LOADS RECOMMENDED BY CHAIN
MANUFACTURERS

(For rigging operations using about one-half these values.)

Nominal size	Iron chain, lb	Acco End-weldur		Taylor		Herc-Alloy, lb	Yale & Towne chain hoist, lb
		No. 125, lb	No. 85, lb	Flash-Alloy, lb	Heavi-Lift, lb		
$\frac{1}{4}$	1,060	2,750	2,150	2,750	2,100		
$\frac{3}{8}$	2,385	6,600	4,275	6,600	4,300	6,600	2,000
$\frac{1}{2}$	4,240	11,250	7,000	11,125	7,000	11,250	4,000
$\frac{5}{8}$	6,630	16,500	10,125	16,500	10,100	16,500	
$\frac{3}{4}$	9,540	23,000	14,000	23,000	14,000	23,000	
$\frac{7}{8}$	12,960	28,750	19,125	28,700	19,100	28,750	
1	16,950	38,750	24,250	38,750	24,250	38,750	
$1\frac{1}{8}$	20,040	44,500	30,250	44,500			
$1\frac{1}{4}$	24,750	57,500	37,500	57,500		57,500	
Safety factor relative to	Proof load	Proof load	Proof load	Proof load	Proof load	Proof load	Breaking load
	2	2	2	2	2	1.35	5

of each chain, including the date of purchase, make, grade of steel,
safe working strength (when new), length for five links (exact), etc.

At his periodic inspection the inspector must carefully examine
each link in the chain. Remember the proverb about a chain
being only as strong as its weakest link. If you overlook one link,
that one may be the defective link. It will be necessary to wipe
off most of the grease to make the inspection properly.

Look for links that may be elongated owing to overloading. If
the links bind, condemn the chain. A free-hanging chain that is
not perfectly straight is indicative of binding links. If there is

any question as to the stretching of the chain, measure five links and compare the present length with the original length as indicated in the record for this particular chain. Condemn the chain if the stretch exceeds 10 per cent. (See Chap. XVIII for test gauge.)

Watch for links that may have become slightly bent when the sling was used to pick up a load with sharp steel or iron edges. The presence of a crack, however fine, particularly at the weld, is sufficient cause for having the link removed from the chain.

If gouges or cuts are observed in the links, they should be very carefully inspected. If the cross section of the link is materially reduced, that link should be cut open and removed from the chain. Watch all sharp nicks and cuts, as it is in these spots that cracks will usually start.

Any link that shows evidence of a crack should be removed from the chain without hesitation. Where a crack is suspected, the link may be soaked in thin oil and then wiped dry. A coating of powdered chalk or other white material is applied to the surface and allowed to remain there for several hours. If a crack exists, the oil pocketed in it will be drawn out by capillary action and will noticeably discolor the white coating.

Small dents similar to peen marks on the surface of the links or even a bright or polished surface usually indicate that the chain has been work-hardened or fatigued.

With the chain lying slack on the floor observe the wear, if any, where the links bear on each other. Figure 1, which has been plotted from data contained in the American Standards Association Safety Code for Cranes, Hoists, and Derricks, shows the reduction in working load on chain corresponding to different amounts of wear. Enter the chart at the bottom from the nominal size of the chain (the actual size is very slightly larger). Also enter from the left side of the chart at the worn dimension, and observe where the lines intersect. If the intersection is below the lowest line, the chain should be removed from service at once.

Strange as it may seem, there is relatively little difference between the strength of a new or worn chain having the same B dimension. For instance, a $\frac{3}{4}$-in. chain that is worn down to $\frac{5}{8}$ in. has approximately the same strength as a new $\frac{5}{8}$-in. chain, but the limit of wear in this case is 25 per cent reduction from A to B dimension. If the chain is taut, it may be desirable to caliper the two links as they bear on each other, then divide by 2 to obtain the B dimension.

Lifting of a fin at the weld is evidence of severe overloading.
If a defective link is observed it should be marked with chalk or
tagged, and the chain returned to the manufacturer.   Do not
attempt to repair it yourself.   Never use a bolt to join the two

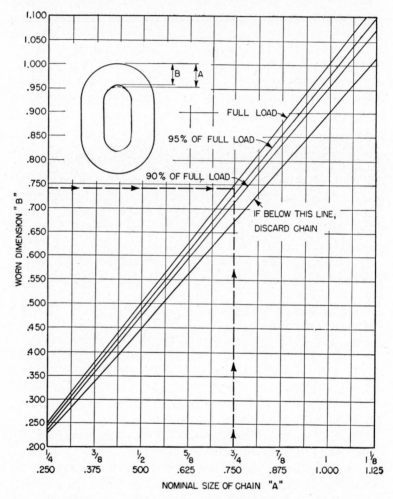

FIG. 1.   Reduction in the strength of chain due to wear.

parts of a broken chain.   Repair links are available for repairing
broken chains; however, their use is discouraged because their
strength is not equal to the strength of the original links (Table
II), and the strength of the entire chain is thus greatly reduced.

Remember that a chain is only as strong as its weakest link. Also, do not attempt to shorten one leg of bridle sling by twisting it. This also causes a serious reduction in strength.

In Table I the sketches of the links will assist in identifying the make and the grade of steel in the chain. For instance, iron chain has a relatively smooth link with an inconspicuous weld. Acco Endweldur chain has the weld at the end of the link, being quite prominent on the inside of the link. The figures 85 or 125 are stamped on the side of each link.

Taylor Flash-Alloy and Heavi-Lift chains of larger size have welds on both sides of each link, being prominent on the inside of the link. Herc-Alloy chain has a weld on one side of the link,

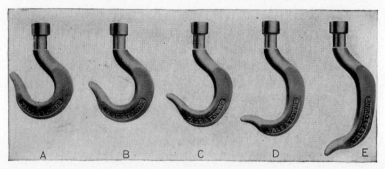

FIG. 2. If the inside contour of the hook is not a true arc of a circle the hook has been badly overloaded. Photographs show a 1-ton (2,000 lb) hook subjected to various degrees of overloading. (*A*) 4,000 lb. (*B*) 5,800 lb. (*C*) 6,800 lb. (*D*) 7,400 lb. (*E*) ? lb. (*Yale & Towne Mfg. Co.*)

noticeable on the inside of the link. Yale & Towne chain for chain hoists has the weld on one side of the link but is a uniform cylindrical bulge.

The ring of a sling chain may be considered as a large link and should be inspected as is the chain itself.

The hook should be the weakest part of any crane, hoist, or sling. It seldom, if ever, breaks, but it may fail by straightening out and finally releasing the load. A distorted hook is prima-facie evidence of overloading, for as shown in Fig. 2, 100 per cent overload does not change the contour of the hook. One hundred and ninety per cent overload is hardly noticeable. Yet it is not uncommon to find hooks spread as shown in the third illustration in Fig. 2c. Reforging and/or annealing of spread hooks should not be permitted. When inspecting a hook pay particular attention to the small radius fillets at the neck.

TABLE II. SAFE LOADS ON CHAIN REPAIR LINKS*

(Compared with Strength of Acco-Endweldur Chain)

Link Size, in.	Safe Load on Link, lb†	Safe Load on Chain, lb
$\frac{1}{4}$	1,115	2,750
$\frac{3}{8}$	2,240	6,600
$\frac{1}{2}$	3,300	11,250
$\frac{5}{8}$	5,000	16,500
$\frac{3}{4}$	7,080	23,000
$\frac{7}{8}$	9,600	28,750
1	12,400	38,750
$1\frac{1}{8}$	15,600	44,500
$1\frac{1}{4}$	20,620	57,500

* U.S. Navy, Bureau of Ships.
† Chain strength from Table I. Safety factor = 5.

After many years of active service all load hooks should be carefully inspected. Hooks of 2 tons capacity or less may be economically junked and replaced by new hooks. Those of 3 to about 40 tons capacity should be dipped in or coated with thin lubricating oil and then wiped thoroughly dry. Then the hook should be painted with a coat of whitewash and allowed to dry. Flaws or cracks will be indicated by discoloration of the whitewash as the oil is leached out of the tiny surface defects. Hooks of 50 tons and larger should be scientifically inspected by using a commercial Magniflux method.

The strength of eyebolts to which a sling can be attached is also of utmost importance, as it is influenced greatly by the direction of pull, as indicated in Table III. When bridle slings are used, an angular pull is always developed in the eyebolts unless a spreader bar is used as part of the sling. See Chap. V for the safe working load on slings at various angles. Unless the pull is in line with the shank of the eyebolt, the strength is very greatly reduced. In the upper portion of Table III are given the strengths of ordinary drop-forged steel eyebolts of the type generally used in rigging operations. The lower portion of Table III contains data developed by Mr. M. F. Biancardi, Manager of the Safety Services Section of Allis-Chalmers Manufacturing Co., and published in the *National Safety News*, relative to a new shoulder-type eyebolt which he designed. When screwed in *tightly*, with the shoulder bearing firmly against the boss or pad on the casting, this type of eyebolt has an angular strength several times as great as the conventional eyebolt.

TABLE III. SAFE LOADS ON EYEBOLTS

Ordinary Drop Forged Steel Eyebolts

Size		PULL	
$\frac{1}{2}''$	1,100 lb	50 lb	40 lb
$\frac{5}{8}''$	1,800 lb	90 lb	65 lb
$\frac{3}{4}''$	2,800 lb	135 lb	100 lb
$\frac{7}{8}''$	3,900 lb	210 lb	150 lb
$1''$	5,100 lb	280 lb	210 lb
$1\frac{1}{4}''$	8,400 lb	500 lb	370 lb
$1\frac{1}{2}''$	12,200 lb	770 lb	575 lb
$1\frac{3}{4}''$	16,500 lb	1,080 lb	800 lb
$2''$	21,800 lb	1,440 lb	1,140 lb

Drop Forged Steel Shoulder-type Eyebolts

Size		PULL	
$\frac{1}{4}''$	300 lb	30 lb	40 lb
$\frac{1}{2}''$	1,300 lb	140 lb	150 lb
$\frac{3}{4}''$	3,000 lb	250 lb	300 lb
$1''$	6,000 lb	500 lb	600 lb
$1\frac{1}{4}''$	9,000 lb	800 lb	900 lb
$1\frac{1}{2}''$	13,000 lb	1,200 lb	1,300 lb
$2''$	23,000 lb	2,100 lb	2,300 lb
$2\frac{1}{2}''$	37,000 lb	3,800 lb	4,300 lb

TABLE IV. SAFE LOADS ON SHACKLES*

Shank, in.	Safe load, lb†	Pin, in.	Inside width, in.	Inside length, in.
$\frac{1}{2}$	2,830	$\frac{5}{8}$	$\frac{13}{16}$	$1\frac{1}{16}$
$\frac{9}{16}$	3,580	$\frac{5}{8}$	$\frac{7}{8}$	$1\frac{13}{16}$
$\frac{5}{8}$	4,420	$\frac{3}{4}$	$1\frac{1}{16}$	2
$\frac{3}{4}$	6,360	$\frac{7}{8}$	$1\frac{1}{4}$	$2\frac{5}{16}$
$\frac{7}{8}$	8,650	1	$1\frac{7}{16}$	$2\frac{7}{8}$
1	11,310	$1\frac{1}{8}$	$1\frac{11}{16}$	$3\frac{1}{4}$
$1\frac{1}{8}$	13,360	$1\frac{1}{4}$	$1\frac{13}{16}$	$3\frac{1}{2}$
$1\frac{1}{4}$	16,500	$1\frac{3}{8}$	2	$3\frac{11}{16}$
$1\frac{3}{8}$	19,960	$1\frac{1}{2}$	$2\frac{1}{4}$	$4\frac{1}{2}$
$1\frac{1}{2}$	23,740	$1\frac{5}{8}$	$2\frac{3}{8}$	5
$1\frac{5}{8}$	27,900	$1\frac{3}{4}$	$2\frac{5}{8}$	$5\frac{1}{4}$
$1\frac{3}{4}$	32,320	2	$2\frac{7}{8}$	$5\frac{7}{8}$
2	42,220	$2\frac{1}{4}$	$3\frac{1}{4}$	$6\frac{3}{4}$

* U.S. Navy, Bureau of Ships.
† Safety factor = 5.

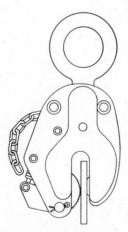

FIG. 3. Automatic clamp for lifting steel plates on edge. (*J. C. Renfroe & Sons.*)

To attach slings, either of chain or wire rope, to the eyebolts on the load to be lifted, it is sometimes the practice to use a standard shackle. Table IV gives the safe working loads on shackles when a safety factor of 5 is used.

For handling steel plates in a vertical position a heavy-duty automatic clamp, as shown in Fig. 3, is recommended. Somewhat similar clamps are available for picking up plates horizontally. There are an unlimited number of types of special sling hooks (Fig. 4), such as for picking up barrels, cases, coils of wire, etc.

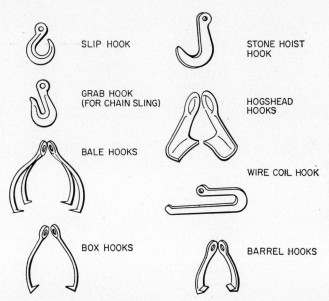

SLIP HOOK

STONE HOIST HOOK

GRAB HOOK (FOR CHAIN SLING)

HOGSHEAD HOOKS

BALE HOOKS

WIRE COIL HOOK

BOX HOOKS

BARREL HOOKS

Fig. 4. Special hooks for hoisting slings. (*U.S. Navy, Bureau of Ships.*)

Table V gives the safe loads for turnbuckles, such as for use on guys or on adjustable sling legs. Table VI shows the safe loads

TABLE V. SAFE LOADS ON TURNBUCKLES*

Size, in.	Load, lb†
$\frac{1}{2}$	1,500
$\frac{5}{8}$	2,400
$\frac{3}{4}$	3,600
$\frac{7}{8}$	5,000
1	6,600
$1\frac{1}{8}$	8,200
$1\frac{1}{4}$	10,600
$1\frac{1}{2}$	15,600
$1\frac{3}{4}$	21,000
2	27,600
$2\frac{1}{4}$	36,200

* American Steel & Wire Co.
† Safety factor = 5.

Table VI. Safe Loads on Sling Hooks*

Depth at Load Point, in.	Safe Load, lb
$1\frac{1}{16}$	2,000
$1\frac{3}{8}$	3,000
$1\frac{7}{16}$	4,000
$1\frac{5}{8}$	5,000
$1\frac{7}{8}$	6,000
2	7,500
$2\frac{1}{2}$	12,000
$2\frac{3}{4}$	15,000
$3\frac{1}{8}$	20,000

* American Steel & Wire Co.

Table VII. Safe Loads on Crane and Derrick Hooks

Diameter of Neck, in.	Safe Load, tons
$1\frac{7}{8}$	5
$2\frac{1}{2}$	$7\frac{1}{2}$
$2\frac{5}{8}$	10
3	15
$3\frac{1}{2}$	20
$3\frac{7}{8}$	25
$4\frac{1}{4}$	30
$4\frac{1}{2}$	35
$4\frac{3}{4}$	40
$5\frac{1}{4}$	50

for sling hooks, based upon the depth of metal in the hook directly below the point where the load is applied. Table VII may be used for checking the main hoisting hook, such as the hook attached to the load block of a derrick.

CHAPTER V

SLINGS

Adequate consideration may be given by the designing engineer to the choice and installation of the component parts of the hoisting tackle on a crane or derrick, including the hoisting cables, their attachments, sheaves, pins, blocks, and hooks, but his effort may be largely nullified should the riggers who eventually operate the equipment use improper slings or use the correct slings in an improper manner.

Manila rope and iron and steel chains have been largely superseded by steel wire rope for the hoisting tackle on such equipment, with resultant increase in safety and with the ability to lift much greater loads. It must be kept in mind that manila rope and chain both have their proper places on rigging jobs; manila rope is used for lifting comparatively light loads and on temporary work, while chains are frequently used in foundries where exposed to high temperatures. But on permanent installations and in heavy-duty operations they have their limitations. Manila rope deteriorates rapidly when exposed to the weather, and after such exposure its actual strength is very difficult to estimate.

Chain has much greater resistance to the effects of exposure to the elements but has other shortcomings. In hand-forged chains there is ever present the human factor in the form of the welder who may or may not have made a perfect weld. Electrically welded steel chain is considerably more uniform in quality, but even its safety depends upon the proverbial "weakest link," as the failure of any one link will allow the load to drop.

Wire hoisting rope, on the other hand, is frequently of the 6×19 construction; that is, it consists of six strands, each containing 19 wires, laid around a hemp-rope center. Unlike the chain, failure of one of the component parts (a wire) will result in a reduction in strength of less than 1 per cent, which is insignificant. This is the primary reason why wire rope has been almost universally adopted for hoisting operations.

But properly designing a piece of hoisting equipment to lift a given load on the hook safely is not the only problem. That load

95

must be secured to the hook, and the attachment must be of adequate strength to lift the load safely. Thus is presented the problem of properly designing and using slings, about which this chapter is written.

Just as in the case of hoisting tackle, various materials can be used for slings. For the very lightest loads an endless manila rope, looped into a noose around the object to be lifted, is satisfactory. Such a sling is inexpensive, of light weight, flexible, and easy to handle. It can be bent around the comparatively sharp edges of boxes or crates but should be padded when passing over the sharp machined edges of metal parts. Even on some heavier jobs, such as handling of steel shafts which must not be scratched or burred, manila-rope slings have found favor.

Coil chains have a limited application as slings, being used principally in foundries where they are exposed to high temperatures and where they are to be used to pick up rough castings that would quickly destroy other types of slings. Like manila rope, a chain should preferably be padded where bearing on sharp edges of metal parts; otherwise some of the links may be subjected to severe bending stresses for which they were not designed. Also, the chain may bruise or otherwise damage the edges of the piece being lifted.

For special work, slings are sometimes made of other materials such as roller chains, leather, and cotton webbing, but by far the greatest number of slings in use today are of wire rope. From the safety viewpoint, wire-rope slings possess the same points of superiority as do wire ropes for hoisting tackle mentioned above. But even wire-rope slings must be carefully chosen for the service in which they are to be placed, as they also have certain limitations. Like manila-rope and chain slings, wire-rope slings should be padded where being bent over sharp edges of the object to be lifted. Although the edge may be of soft material such as a wood crate or skid, which will not cut the sling, nevertheless the individual wires being bent sharply will be highly stressed, possibly beyond the elastic limit, even before the live load is applied.

Wire-rope slings must not be allowed to become kinked, as this also causes severe bending stress and, in addition, the strands that have been displaced are subject to unequal distribution of the live load, some strands taking more than their normal load, in addition to the stress produced by the sharp bending. Even though an attempt is made to remove a kink, the damage done to the rope is usually permanent. This accounts for many sling failures. Most wire-rope slings, like hoisting cables, are of 6 × 19 filler wire construction and of improved plow steel grade. Slings of $\frac{1}{2}$-in. rope and larger should be made from wire rope with an independent wire-rope center. This will reduce the probability of the rope being crushed in service.

The very large sizes are often made of 6 × 37 or 6 × 6 × 19 construction to increase the flexibility. All slings that have at one time or another been bent sharply become "cranky"; that is, they have developed permanent deflections, so that, when the strain is relieved, they will snarl in some unpredictable manner. This is very annoying, to say the least, and it means continual vigilance to keep them from developing kinks when taking a strain preparatory to lifting a load. Slings made of preformed rope behave somewhat better than those made in the conventional manner and also have the advantage that broken wires will not so readily wicker out and present a hand hazard to the rigger, but rather tend to lay "dead" in the original position.

FIG. 1. Endless sling or grommet.

The term "sling" includes a wide variety of designs. Perhaps the simplest is the grommet or endless type (Fig. 1). If of wire rope, it is made by using six complete loops of a single strand laid or twisted around itself on each successive loop. The ends are then tucked into the space where a short length of the hemp center has been removed, such as is done in making a long splice. This type of sling is usually formed into a noose known as a "choker hitch," or "anchor hitch," which is slipped around the object to be lifted, and the free bight is placed on the crane or derrick hook. When a strain is taken on the sling, it crowds in and squeezes the load, but to assure that it grips the load firmly, it is good practice to strike the bight that bears on the hauling part with a 2 × 4 so as to make three 120-deg angles between the component parts of the sling at the bight.

For handling miscellaneous loads the rigger should have available four single slings, six shackles, four hooks, and a ring. With

Fig. 2. Component parts for a single sling, choker sling, or basket sling. (*American Chain & Cable Co.*)

Fig. 3. Basket hitch. If the load is prevented from tilting, such as by the use of two hoists or cranes, this hitch is suitable for lifting shafts, boiler drums, tanks, etc. (*American Chain & Cable Co.*)

Fig. 4. A choker sling gets a viselike grip on the load. (*American Chain & Cable Co.*).

this equipment he will be able to lift almost any type of load, provided it is not too heavy or too large for the particular slings.

Having one single sling and the accessories shown in Fig. 2, he can lift many loads safely with or without these accessories. It can be used as a basket sling (Fig. 3) for lifting certain types of loads. The middle of the sling is placed through the load to be lifted, and both spliced eyes are placed on the crane or hoist hook.

Or the sling can be wrapped around the load, one eye threaded through the other eye and then placed on the crane hook. This is known as a "single choker sling" (Fig. 4).

If it is not convenient to slip the noose on and off the end of the load,

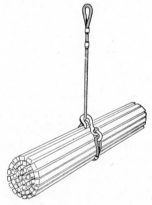

Fig. 5. Choker sling can also be used with a hook where it is more convenient. (*American Chain & Cable Co.*)

the shackle can be used to attach the hook to the lower eye. The sling is then passed around the load and hooked into itself to form a choker (Fig. 5). In order to reduce wear and sharp bending where the hook bears on the body of the sling, a special forging is sometimes threaded onto the sling. The eye is made so as to cause a minimum bending of the rope, and it has a hook onto which the eye of the sling is placed. The choker sling can be used to lift one

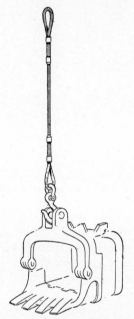

FIG. 6. A single-leg sling can be used for a direct lift where the hoist hook is too big to engage the lifting bolt or pin on the load. (*American Chain & Cable Co.*)

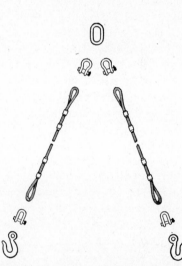

FIG. 7. Component parts for a two-leg bridle sling, two basket slings, two choker slings, etc. (*American Chain & Cable Co.*)

or a bundle of pipes, rods, etc., but such loads should be carefully balanced to avoid the possibility of one of the interior pieces slipping out of the bundle.

For lifting such loads as motors which have lifting eyebolts, the single-leg sling can be used straight (Fig. 6).

Two of these slings can be used as a pair being known as a two-leg sling (Fig. 7). Without the accessories, they can be used as a double basket sling (Fig. 8) for lifting almost any type of load, such as stacks of sheets or plates. Or they can be used as a double choker for handling an irregular-shape load (Fig. 9). With the shackles and hooks attached, the slings can also be used as a double choker (Fig. 10) for picking up any long round, square, or irregular-shape load.   Figure 11 shows these slings used as a bridle sling for hoisting large pipe or for handling other objects having lifting bolts or other means of attachment.

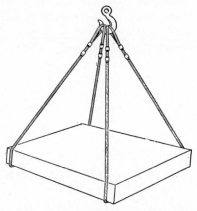

Four of the single slings can also be used as a unit (Fig. 12). Without the accessories they can be used as a four-leg bridle sling for lifting an object that has the necessary lifting lugs or handles (Fig. 13), or with the hooks attached they can be used to attach to eyebolts or holes in the object (Fig. 14).

FIG. 8.   In using the double basket sling, watch that one part does not tend to slip along the load and allow it to tilt and drop.   (*American Chain & Cable Co.*)

With the hooks removed but with the eyes of the slings joined by the shackles, the slings can be lengthened as required for lifting larger loads (Fig. 15).   If still greater length of sling is required, two additional slings (Fig. 16) can be used in conjunction with the four-leg sling to form a double basket.

FIG. 9.   Double choker sling with spliced eyes.   (*American Chain & Cable Co.*)

When shackles are used as indicated, short pieces of tubing or pipe should be placed on the shackle pins to increase their diameter and thus avoid bending the sling eye too sharply.

For lifting sturdy shipping cases, a special sling (Fig. 17) will be very useful.   By means of the small sheaves, the sling causes the grab hooks to be forced into the wood.   With the use of a similar sling but with blunt hooks, drums or barrels can be handled with ease and safety.

The special turnbuckle sling shown in Fig. 18 is used for lifting

turbine casings, which have to be raised on an even keel in order
not to foul the turbine blading.

In recent years a type of wire-rope sling has been developed that
is replacing many of the conventional type using the ordinary
straight cable. This new type is the interwoven or braided sling
made up of a number of smaller size wire ropes, usually 6 but
occasionally as many as 48 parts. Figure 19a shows a Macwhyte

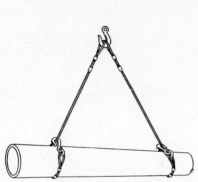

Fig. 10. Double choker sling with hooks
attached for more readily attaching and de-
taching. (*American Chain & Cable Co.*)

Fig. 11. The bridle sling.
(*American Chain & Cable Co.*)

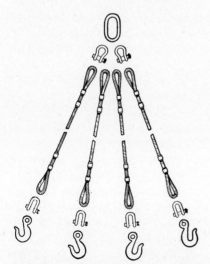

Fig. 12. Component parts for making
any of the above illustrated slings or for
four-leg bridle sling or double basket sling.
(*American Chain & Cable Co.*)

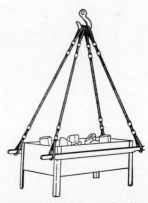

Fig. 13. Four-leg bridle sling
with spliced eyes. (*American
Chain & Cable Co.*)

Drew sling composed of one piece of wire rope spliced endless before braiding into a flat multiple-part body. In Fig. 19*b* is shown an Atlas sling, which has a round braided body made from two pieces of wire rope, one right lay and one left lay. The two ropes are spliced endless, folded to secure the required numbers of parts, and then braided. All ropes form a continuous, uniform spiral throughout the entire length of the sling.

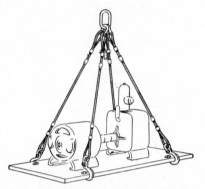

FIG. 14.   Four-leg bridle sling with hooks for use where lifting holes or eyebolts are provided. (*American Chain & Cable Co.*)

Some of the larger sizes are braided around a hemp-rope core, which acts as a cushion and assists in keeping the sling nearly round in cross section. Even on the eight-part slings the spiral braiding gives a much lower modulus of elasticity than a single cable of the same capacity. This greatly reduces the stress on the individual wires when a load is applied rather suddenly, thereby increasing the safety factor. Braided slings are also constructed so as to be almost flat in the general cross section, these being used particularly for basket-type hitches, especially when lifting heavy shafts.

FIG. 15.   Where a large load requires longer slings, the four legs can be used as shown to form a double basket sling. (*American Chain & Cable Co.*)

For the basket hitch, when the diameter of the object hoisted is very large such as a locomotive boiler, a sling made up of a number of parts of wire rope wound back and forth between two heavy-duty thimbles is used. These parts of the rope are laid parallel to each other and are served at frequent intervals to bind them into one unit. Such a sling should never be bent around small diameters or passed over sharp edges of the load.

Of course, all slings should be made of improved plow steel rope, which is usually identified by various trade names and trade markers. All eye splices should be made by experienced splicers

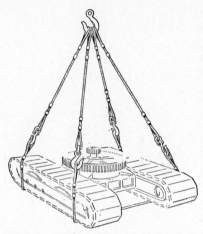

and should be properly served to conceal the sharp protruding ends of the wires, which frequently cause injury to the riggers handling them. While not absolutely necessary, it is good practice to place thimbles in all spliced eyes. Socketing, likewise, should be done only by experienced men. Care should be exercised when socketing not to allow the twist or lay to come out of the rope as it enters the throat of the socket. When not in use, slings should be hung up in an orderly manner on special hooks or brackets so as to keep them as straight as practicable.

FIG. 16. Where a still larger load is to be lifted, a four-leg bridle sling can be made into a large double basket sling by the use of two additional single-leg slings. (*American Chain & Cable Co.*)

Special consideration should be given to the angle that the legs of a sling make to the horizontal, as the lifting capacity is reduced much more rapidly than might be expected as the angle is reduced.

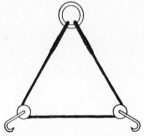

FIG. 17. A special sling for handling small packing cases, etc.

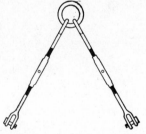

FIG. 18. Where the load to be lifted must be maintained in an absolutely level position, turnbuckles can be made an integral part of the sling.

For a single leg or part of the sling at an angle of 60 deg, the lifting capacity is only 87 per cent of that at 90 deg or vertical, at 45 deg it is 71 per cent, and at 30 deg it is only 50 per cent.

TABLE I. SAFE WORKING LOADS ON VARIOUS TYPES OF SLINGS

Type of sling	Nominal size, in.	Single sling, lb	Choker sling, lb	U sling, lb	Basket sling, lb	Total load on two-leg slings (For three-leg sling multiply by 1½. For four-leg sling multiply by 2)			Weight per ft (exclusive of hook, ring, thimble, or splice), lb
						60-deg bridle, lb	45-deg bridle, lb	30-deg bridle, lb	
6 × 19 improved plow steel rope (Federal spec. RR–R–571) Factor of safety = 8 Splice efficiency = 80% Rope diameter →	3/8	1,350	1,010	2,700	2,360	2,330	1,910	1,350	0.23
	7/16	1,840	1,380	3,680	3,220	3,180	2,600	1,840	0.31
	1/2	2,420	1,815	4,840	4,240	4,180	3,420	2,420	0.40
	9/16	2,900	2,175	5,800	5,080	5,000	4,110	2,900	0.51
	5/8	3,800	2,850	7,600	6,650	6,570	5,400	3,800	0.63
	3/4	5,260	3,940	10,520	9,200	9,100	7,450	5,260	0.90
	7/8	7,000	5,250	14,000	12,250	12,100	9,900	7,000	1.23
	1	9,000	6,750	18,000	15,750	15,550	12,750	9,000	1.60
	1⅛	11,200	8,400	22,400	19,600	19,400	15,900	11,200	2.03
	1¼	13,800	10,350	27,600	24,200	23,900	19,550	13,800	2.50
Iron crane chain (ASTM spec. A 56–39) Factor of safety = 5 Stock diameter →	3/8	1,710	1,280	3,420	3,000	2,970	2,420	1,710	1.66
	1/2	2,845	2,130	5,690	4,980	4,940	4,030	2,845	2.75
	5/8	4,380	3,280	8,760	7,680	7,600	6,200	4,380	4.30
	3/4	6,415	4,820	12,830	12,200	11,150	9,100	6,415	6.15
	7/8	8,850	6,630	17,700	15,500	15,350	12,680	8,850	8.20
	1	11,750	8,800	23,550	20,600	20,400	16,650	11,775	10.45
	1⅛	15,350	11,500	30,700	26,900	26,650	21,700	15,350	13.10
	1¼	19,250	14,400	38,500	33,700	33,500	27,250	19,250	16.00

Refer to the figures given in Fig. 1, Chap. IV, and multiply by the use factors indicated below

Alloy-steel chain							
Factor of safety = 5 ±							
Stock diameter →							
$9/32$							0.80
$5/16$							0.92
$3/8$							1.33
$7/16$							1.80
$1/2$							2.36
$9/16$							3.00
$5/8$							3.80
$3/4$							5.45
$7/8$							7.50
8-part braided wire strand sling							
Factor of safety = 5							
Splice efficiency = 100%							
Strand diameter →							
$1/8$	1,540	1,150	3,080	2,700	2,670	1,630	1,540
$3/16$	3,420	2,560	6,840	5,990	5,920	4,850	3,420
$1/4$	6,500	4,870	13,000	11,380	11,300	9,220	6,500
$5/16$	10,800	8,100	21,600	18,850	18,700	15,300	10,800
$3/8$	14,120	10,600	28,240	24,650	24,400	20,000	14,120
$7/16$	18,820	14,150	37,640	33,000	32,600	26,600	18,820
$1/2$	24,200	18,150	48,400	42,300	41,900	34,300	24,200
$9/16$	30,240	22,700	60,480	53,000	52,300	42,900	30,240
3-strand manila rope (Federal Spec. T–R–601)							
Factor of safety = 10							
Splice efficiency = 80%							
Rope diameter →							
$1/2$	210	160	420	370	365	300	210
$3/4$	435	325	870	760	740	615	435
1	720	540	1,440	1,260	1,250	1,020	720
$1\frac{1}{4}$	1,080	810	2,160	1,900	1,880	1,530	1,080
$1\frac{1}{2}$	1,480	1,110	2,960	2,600	2,570	2,100	1,480
Use factor	1.000	0.750	2.000	1.750	1.732	1.414	1.000

Table I gives the safe working loads for the conventional wire-rope slings, iron-chain slings, alloy steel slings, and braided slings. For manila-rope slings a safety factor of 10 is used.   The loads are given for two-leg bridle slings at various angles.   For three-way and four-way bridle slings these loads can be increased 50 per cent and one hundred per cent, respectively.   According to the ASA Safety Code for Cranes, Hoists, and Derricks, chain slings should not be loaded to more than one-half of the proof test load.

a   *b*

Fig. 19.   (*a*) Six-part flat-braided sling.   (*b*) Eight-part round-braided sling. (*Macwhyte Company.*)

Of course, slings should be inspected periodically and condemned when found in an unsafe condition.   A few broken wires do not perceptibly weaken the sling unless located near the throat of a socket, in which case they may indicate fatigue of the metal of the rope.   If crushed not too severely the strands can be hammered back into shape by means of a wood mallet and a wood block. Slings, regardless of type, should be protected from corrosion and from contact with injurious chemicals.   With reasonable care a sling should last indefinitely and should be one of the safest parts of the hoisting equipment.

Synthetic-fiber slings (Fig. 20) are available for handling accurately machined or ground shafts, motor rotors, bundles of tubing, etc. Table II gives the approximate safe working loads for typical synthetic slings, based on a 1-in. width. For actual safe load multiply figure given by the sling width in inches. Warning: Do not use flat slings as spreaders, such as on a shaft, because in such a position one edge of the material will take

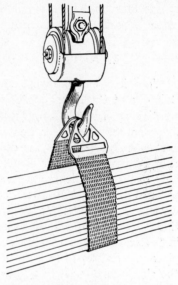

Fig. 20. Synthetic fiber sling lifts machine parts safely and without scratching or otherwise damaging the part. (*Jones & Laughlin Steel Co.*)

Fig. 21. Flat chain mesh sling. (*Cambridge Wire Cloth Co.*)

most or all of the stress. As an example, one may apply considerable tensile stress to a strip of paper, yet if the pull is all at one edge of the paper it will readily tear it.

TABLE II. SAFE LOADS ON SYNTHETIC FIBER SLINGS

Material	When used singly	As choker hitch	As basket hitch
Nylon	1,200 lb/1 in. width	900 lb/1 in. width	2,400 lb/1 in. width
Dacron	800 " "	600 " "	1,600 " "

Chain mesh slings are used for handling heavier loads, particularly loads of bundles of pipes, rods, bars, etc. A flexible flat sling has the ability to hug the load closer and thus prevent any of the material from slipping out during handling.

CHAPTER VI

WOOD FOR STRUCTURAL PURPOSES

Wood is one of the most common materials in use today, and it is one of the oldest materials. Notwithstanding these facts, relatively very little is understood of the behavior of wood under stress by the average carpenter, engineer, or rigger.

It therefore becomes the duty of the engineer or other person charged with inspecting structures or objects made of wood not only to check the depreciation and loss of strength of the material, but also to investigate if the correct species and quality of wood were used originally. It is admitted by authorities on the subject that only extreme care in examining a specimen by a competent inspector can even roughly determine the actual condition of a piece of wood that has been in use for some time and exposed to the elements.

Surface defects are more readily observed, but defects such as decay often destroy the entire heart of a timber without any trace on the surface. Admitting the limitations of even the most experienced inspector, the following data are presented so that the engineer, rigger, or job foreman can make his inspections with somewhat greater accuracy.

While the information contained herein is of a general nature, it has been compiled with particular reference to structural timbers, rigging and shoring timbers, scaffold planks, ladders, etc. It includes some of the basic information relative to the structure of wood and its inherent defects, utilization, and identification.

In order that the reader can more fully understand the references made herein to the parts of the material, its characteristics, defects, infection, infestation, etc., it is important that he fully understand the various definitions. Hence, the following list of definitions should be carefully studied.

Annual Rings. The concentric but often irregular, recurring rings that appear in whole or in part upon the cross section of any native wood, whether it be in the form of a tree, log, timber, plank, or finished wood part. Each complete annual ring consists of an inner band of low-density wood (springwood) and an outer band of higher density wood (summerwood).

Bastard Sawn. Lumber so cut that the annual rings make an angle of between 30 and 60 deg with the faces of the piece.

Blue Stain. A bluish or grayish discoloration of sap wood of certain species caused by the growth of certain moldlike fungi on the surface and in the interior of the wood.

Boxed Heart. A timber is said to have boxed heart when the pith is located within the piece.

Brashness. A brittle condition of wood characterized by a more or less abrupt failure across the grain, instead of a tendency to splinter, when broken.

Brown Stain. A chemical discoloration of white pine and yellow pine, apparently due to oxidation and accumulation of extractives under certain conditions during the drying of various species.

Bruise. An injury to the wood at or near the surface caused by its having been struck by a hammer or other object. Such spots are vulnerable to decay.

Check. A lengthwise separation or split in the wood that occurs radially, or normal to the annual rings, as seen on the cross section. This is the result of uneven shrinkage of the wood.

Compression Failure. The buckling of fibers due to excessive compression along the grain as a result of flexure or end compression of the piece. It may occur as a result of severe bending of a tree by wind, ice, and snow or by stressing wood structural members beyond the proportional limit.

Compression Wood. An abnormal growth occurring in conifers and characterized by relatively wide annual rings, usually eccentric, and a comparatively large proportion of summerwood (usually 50 per cent or more), which merges into the springwood without a marked contrast in color.

Conifer. A tree bearing seed cones, usually an evergreen, the wood from which is known as "softwood."

Cross Break. The separation of wood cells across the grain, such as may be due to tension resulting from unusual shrinkage or mechanical stresses.

Cross Grain. The grain, or direction of the cells in a piece of wood, that does not run parallel with the axis of the piece.

Burl. A local disturbance in the grain, usually associated with knots, or undeveloped bands produced by the healing of wounds during the life of the tree.

Curly Grain. When the fibers are irregularly distorted as in maple, birch, and other species. This may be several inches long as observed on the tangential surface.

Diagonal Grain. When the fibers run diagonally across the piece as a result of the latter having been sawed at an angle across the annual rings. It may be observed on the radial and occasionally on the tangential face.

Dip Grain. A wave or undulation such as occurs around a knot, pitch pocket, or other defect.

Interlocked Grain. An instance in which spiral grain occurs in one direction for a number of years, then in the other direction for a number of years. Observed when split on a radial line.

Spiral Grain. When the fibers take a spiral or winding course in the tree or log. May be detected on the flat grain surface.

Cross Section. The section of a tree, at right angle to the pith or axis, as observed on the end view of a log or piece of wood cut from it.

Decay or Rot. Disintegration of the wood substance due to the action of wood-destroying fungi.

Edge Grain, Vertical Grain, Rift Grain, Comb Grain, or Quarter Sawed. Lumber that has been sawed so that the annual rings run at an angle of 45 deg or more with the wide face of the piece (see Fig. 1).

Flat Grain. Lumber in which the annual rings form an angle of less than 45 deg with the wide face of the piece. Also called slash grain, plain sawed or tangential cut (see Fig. 1).

Heartwood. The inner and usually darker portion of the cross section of a tree, all cells of which are lifeless cases and serve only the mechanical function of keeping the tree from breaking under its own weight and from the force of the wind.

Hollow Heart. A cavity in the heart of a log resulting from decay.

Honeycombing. Open checks in the interior of the piece of lumber, often not visible on the surface.

Knots. Cross sections of branches or limbs of a tree within the lumber such as cut from the trunk of a tree. Knots may be round or spiked, sound or unsound, tight or loose, decayed, intergrown, encased, not firm, pith knot, or hollow knot.

Live Timber. Timber cut from a tree that was living at the time of cutting.

Low-density Wood. Wood that is exceptionally light in weight for its species, due usually to abnormal growth conditions. It is frequently referred to as "brash wood" and breaks with a splinterless fracture.

Pitch. A poorly defined accumulation of resin in the wood cells in a more or less irregular patch.

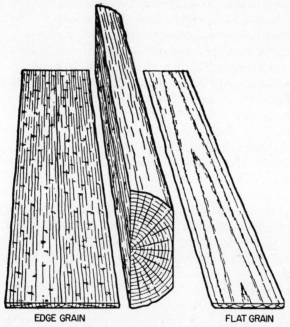

EDGE GRAIN FLAT GRAIN

Fig. 1. How edge-grain and flat-grain boards are cut from a log. (*Forest Products Laboratory.*)

Pitch Pocket. A well-defined opening between annual rings, containing more or less pitch.

Pith. The soft core at the center of the trunk or limb of a tree or log.

Pores. The hollow cells or vessels of which the wood is composed, usually being larger and thin-walled in the springwood, and smaller and heavy-walled in the summerwood, as observed on the cross section under a microscope. These cells, tubes, pores, or tracheids run vertically in the tree and conduct the sap up and down the tree.

Radial Face. That face of a piece of wood which extends in a generally radial direction in the tree. This surface cuts the annual rings at nearly a right angle and usually presents edge grain.

Rays. The tiers of cells extending radially in the tree. They are seen as bands on the radial face of a piece of wood and as dashes on the tangential face. The entire radial face will be seen almost covered with these tiny structures, which appear as fine to conspicuous short lines.

Red Heart. The incipient stage of a destructive heart rot in certain coniferous trees caused by *Fomes pini.* It is characterized by an abnormal pink to purplish red or brownish color in the heartwood.

Resin Ducts. Very small irregular openings. These are visible on all three sections of a tree, but sometimes appear as fine pin scratches on the tangential section. They occur normally only in pines, spruces, Douglas fir, larch, and tamarack.

Sap. All the fluids in a tree.

Sapwood. The zone of light-colored wood near the bark on the cross section of a tree, about 1 to 3 in. wide and containing 5 to 50 annual rings. The cells in the sapwood are active and store up starch and otherwise assist in the life processes of the tree, although only the last or outer layer of cells forms the growing part and true life of the tree.

Second Growth. Timber that has grown after the removal of all or a large portion of the previous stand, whether by cutting, fire, wind, or other agency. Second-growth material is frequently of rapid growth during its early life.

Shake. A lengthwise separation of the wood that occurs usually between and parallel to the annual rings. A through shake is one extending between two opposite or adjacent faces of a piece of wood.

Sound Wood. Wood that has not been affected by decay.

Split. A lengthwise separation of the wood, extending from one surface through the piece to the opposite or adjacent surface.

Springwood. The inner portion of the annual ring, which is grown in the spring season. This portion of the ring is less dense than the summerwood.

Stain. A discoloration occurring on or in lumber of any color other than the natural color of the piece.

Summerwood. The outer portion of the annual ring, which is grown in the summer season. This portion of the ring is denser than the springwood.

Tangential Surface. That face of a piece of wood which is tangent to the annual rings in the piece, usually showing flat grain.

Wane. Bark or lack of wood on the edge or corner of the piece.

Warp. Variation of the piece from a straight plane surface:

  Bow. Deviation flatways from a straight line.

  Crook. Deviation edgeways from a straight line.

  Cup. Deviation crossways from a straight line.

  Twist. Deviation spirally from a straight line.

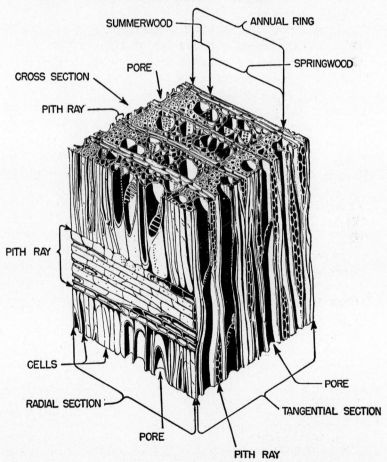

FIG. 2. A $\frac{1}{32}$-in. cube of hardwood greatly magnified. (*Forest Products Laboratory.*)

STRUCTURE OF WOOD

To get down to fundamentals, wood in general is composed of millions (yes, in the order of a million per square inch in some species) of tiny cells, running lengthwise in the tree or log. Sandwiched between them at various places and extending in a radial

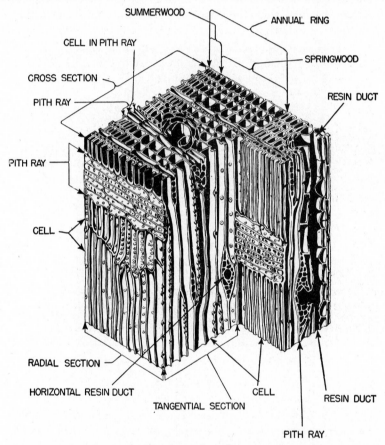

FIG. 3.   A $\frac{1}{32}$-in. cube of softwood greatly magnified.   (*Forest Products Laboratory.*)

direction are long slender groups of cells known as "pith rays," one of the functions of which is to bond the wood together radially (see Figs. 2 and 3).

In the wood of growing trees only that portion, up to about 3 in. thick, beneath the bark contains living cells. This area is known as "sapwood." The darker, central portion of the living tree consists of lifeless cells and is "as dead as a log," this area being known as "heartwood."

Moisture taken in by the roots of the tree ascends through those cells constituting the band of sapwood and travels out through the branches to the leaves. Here it combines with carbon from the air to form food material that descends through the inner bark adjacent to the wood, and part of which develops into wood cells and part into bark.

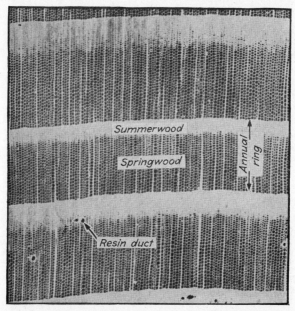

Fig. 4. Cross section of a nonporous wood, magnified about 15 times.

During the winter season the flow of sap is interrupted and tree growth ceases in temperate climates. When the rainy, spring season arrives, tree growth is resumed and large but thin-walled wood cells are produced quite rapidly in order to supply the growing tree with sufficient moisture. During the hot summer months the cells produced are smaller but have thicker walls, thus making this portion heavier and stronger than that developed in the spring season. So each year an additional pair of concentric rings is added to the tree, the lighter part being known as "springwood" and the denser part as "summerwood." Collectively, they are known as one "annual ring."

Domestic woods are divided into two classes: hardwoods and

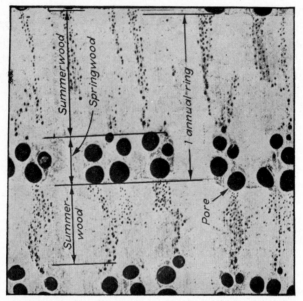

FIG. 5.   Cross section of a ring-porous wood, magnified about 15 times.

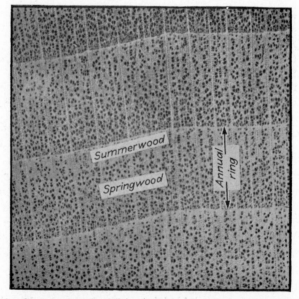

FIG. 6.   Cross section of a diffuse-porous wood, magnified about 15 times.

softwoods. However, all hardwoods are not hard, and all soft-woods are not soft. For instance, longleaf yellow pine is nom-inally a softwood, while poplar and cottonwood are called hard-woods. The answer to this puzzle is that domestic woods are classed botanically rather than according to physical and mechani-cal properties. In other words, the wood from all coniferous trees is known as softwood and is nonporous in structure (see Fig. 4).

Hardwoods, on the other hand, come from trees having broad leaves and are subdivided into two types: ring porous, such as oak, ash, and hickory, in which the pores are arranged in definite rows or rings along the layer of springwood (Fig. 5), and diffuse porous, such as gum, walnut, and maple, in which pores of more uniform size are scattered throughout the annual ring (Fig. 6).

DEFECTS AND OTHER CHARACTERISTICS AFFECTING STRENGTH

Knots. The most common defect in wood is the knot. This may appear either round or elongated, being round where the branch or limb extended at about a right angle to the face of

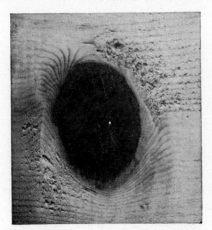

Fig 7.   Sound or intergrown knot.

the piece containing the knot, and elongated or "spike" where the limb extended more nearly parallel to the surface. In judging the strength of a piece containing a knot, it should be borne in mind that the axis of the knot extends to the pith of the tree, the form of the knot being conical with the apex at the pith.

A "sound knot" is one that shows no evidence of decay. An "intergrown knot" (Fig. 7) is one in which the fibers of the knot are intergrown with the fibers of the wood, without reference to soundness. An "encased knot" (Fig. 8) may be surrounded by bark or pitch. A "pin knot" is a tiny knot. A "spike knot" (Fig. 9) is a knot sawed through longitudinally.

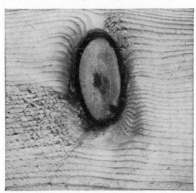

FIG. 8.   Encased knot surrounded by a layer of dead bark.

The weakening effect of a knot is increased by the effect of the cross grain produced in the wood adjacent to the knot. Thus, burls or dip grain, even without the presence of the knot itself, constitute defects in the wood.

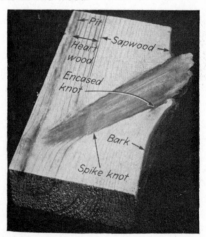

FIG. 9. A spike knot is a longitudinal (or nearly longitudinal) section through a limb or branch of the tree.

Knots are most detrimental when they occur in sections numbered 1 and 2 of Fig. 10.   However, they are somewhat less objectionable in section 2 if they are sound and are securely embedded in the wood so as to transmit the compressive stress.   A sound knot in section 1 is just as weakening as a loose knot or even a knot hole.

Tests on short, deep beams that ordinarily would fail in horizontal shear indicate that sound knots located near the points of maximum horizontal shear do not decrease the strength of the beam as might be expected.   Sound knots may also add to the strength under compression across the grain.

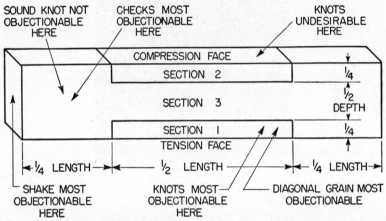

FIG. 10.   Diagram showing the areas in a wood beam in which certain defects are most objectionable.

In long columns, where stiffness rather than direct crushing strength is the controlling factor, the loss of strength due to knots or other defects is relatively small.

In joists or planks knots are measured by their average diameters.   In short or intermediate-length columns the reduction in strength is approximately proportional to the size of the knot. On heavy beams knots on the top and bottom faces are measured between lines drawn tangent to the knot and parallel to the edges of the beam.   On the vertical faces of such beams, knots are measured by their minimum diameters.

Checks.   Perhaps the next, most common defect is the check (Fig. 11).   Checks develop owing to uneven shrinkage when the wood is dried.   The result is the pulling apart or separating of the rows of cells in a radial split, starting usually at the bark or outside of the timber.   The width of a check has comparatively little effect upon the strength of a timber, as a check of almost imperceptible width is as detrimental as a wider one.

Checks are the most harmful if located near the ends of the vertical faces of a beam in the middle half of the height of the beam. At this point the horizontal shearing stress is at its maximum, there being a tendency for a beam to split in half so that the upper

FIG. 11. A check is a radial split in a piece of wood resulting from unequal shrinkage.

half may slip on the lower half, as shown in Fig. 12. The presence of checks at this location reduces the resistance to this horizontal shear and results in the failure of many beams, which otherwise

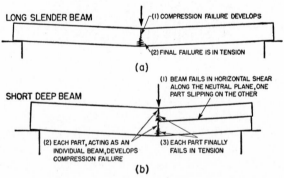

FIG. 12. Typical failures in (a) a long slender beam and (b) a short deep beam.

would carry much greater loads. This is particularly true of beams having a depth greater than one-twelfth of the span. On long slender beams checks do little, if any, harm if the wood is straight grained. Checks on the horizontal faces of a beam do little harm, provided the beam is not subject to lateral bending. Internal checks (on the cross section) often are not apparent on the surface of the piece but may nevertheless greatly reduce its strength.

The effect of checks on the resistance to horizontal shear can be roughly calculated by estimating the actual reduction in area in a horizontal plane at the neutral axis within a distance from the end of the beam equal to three times the depth of the beam.

Checks have little, if any, harmful effect on poles except that they permit the entrance of moisture.

Shakes. Shakes (see definition), like checks, are most objectionable when occurring at the points of maximum horizontal shear. They are measured between vertical lines enclosing the shake, and never should a shake extend to the surface of the piece (see Fig. 13).

Fig. 13. A shake is a separation between successive annual rings.

In poles, shakes should not be permitted if they enclose more than 10 per cent of the area of the cross section. In other words, the diameter of the wood enclosed by the shake should be not more than one-third the pole diameter.

Occasionally an old timber in a structure may be observed with splits running longitudinally on one face and diagonally on the adjacent face, thus posing the question as to whether the grain is straight or spiral. Upon careful examination it will often be found that the diagonal splits are checks, indicating the presence of spiral grain, while the longitudinal splits will be found to be shakes, not checks. Inserting a knife blade into the split and then running it lengthwise should show if it has a tendency to rotate around the pith. If it does, the split is a check; if not, it probably is a shake, assuming that the rings cannot be observed at the end of the timber.

Pitch Pockets. This type of defect in large-size pieces is not too serious, provided the dip grain that accompanies it is not excessive. A number of pitch pockets (see Fig. 14) in the same year's growth in a piece indicates lack of bond of the wood and the probable presence of a shake.

Pitch pockets are characteristic of the pines, the spruces, larch, tamarack, and Douglas fir.

Fig. 14.   Pitch pockets occur only in certain softwoods.

Cross Grain. This is one of the most treacherous defects in a piece of wood, inasmuch as it is often difficult to detect.   Cross grain is a general term that includes diagonal grain, spiral grain, interlocked grain, dip grain, wavy grain, curly grain, etc.

Diagonal grain is the result of sawing a straight piece from a crooked log, or not sawing parallel to the bark in a straight but tapered log, in which case the wood fibers do not run parallel to the edges of the piece.   Diagonal grain may occur on either the edge-grain or flat-grain face of the piece, generally on the former, in which it is usually readily discernible by the angular direction of the annual ring markings.   If it occurs on the flat-grain face, it will be considered in the same class with spiral grain.

Spiral grain is not so easy to detect.   In the growing tree, and consequently in the log cut from it, the wood fibers run in a generally longitudinal direction.   In some trees, however, the fibers run at a slight angle, thus taking a spiral path around the tree.   Such spiral grain may frequently be observed on telephone poles, flagpoles, etc., where it is definitely indicated by the checks.

On the flat-grain face of a piece of wood the markings of the annual rings do *not* indicate the direction of the grain (Fig. 15).   If there are a number of herringbone V's on this face, they indicate diagonal grain, but the angle must be measured on the edge-grain or radial face. If the points of these V's are not in a line parallel to the edge of the piece, spiral grain undoubtedly exists.

In order to demonstrate spiral grain the model shown in Fig. 16 was constructed. It is intended to represent a log with a "beam" cut from it. The annual rings, being for all practical purposes a number of concentric cylinders, intersect the radial face of the beam and produce numerous parallel markings commonly called

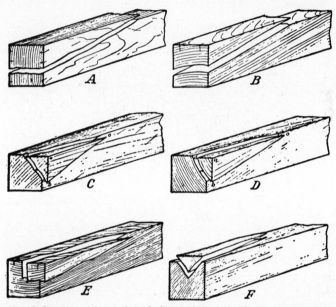

Fig. 15. Indications of spiral and diagonal grain and combinations thereof. Annual rings parallel to edge of piece: (*A*) spiral grain; (*B*) diagonal grain. Annual rings oblique to edge of piece: (*C*) spiral grain; (*D*) diagonal grain. Spiral and diagonal grain in combination: (*E*) rings parallel to edge; (*F*) rings oblique to edge.

"comb grain." These markings, running parallel to the edges of the beam, indicate freedom from *diagonal* grain, and many a rigger or carpenter might consider this beam safe to support a heavy load.

But if we look on the flat-grain face of the piece, we can detect numerous brownish markings not unlike pin scratches. These are resin ducts (they occur normally on the pines, spruces, larch, tamarack, and Douglas fir) which definitely indicate the direction of the grain. On this beam these resin ducts are at an angle of 1:7, which is far beyond the safe angle of 1:15 to 1:20.

Rotating the "log" so as to observe the other side, we note that there are several checks running spirally at the 1:7 angle. If spiral grain exists in the log, of course it must also exist in the beam cut from it, which verifies our interpretation of the resin ducts. To make assurance doubly sure, a large nail was driven into the end of the beam, with the result that a piece split off on the 1:7 angle. This piece is therefore declared unfit for structural purposes, although to the uninformed it would probably be accepted as a perfect specimen.

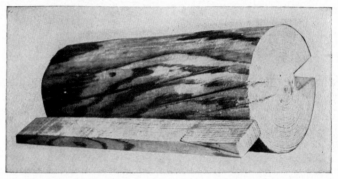

Fig. 16.   Beam cut from a large log.   Spiral grain is indicated by checks and split.

The foregoing will demonstrate the futility of attempting to judge the grain of a piece of wood by the so-called "visible grain" (see Fig. 15).

To determine the direction of the true grain in a piece of wood, the markings, or so-called "visible grain," on the *radial* face are first observed. If these do not run parallel to the edge of the piece, it is indicative of diagonal grain, the angle being measured by the length required to produce a deviation of 1 in. Next, we observe the *tangential* or *flat-grain* surface. Checks, if present, indicate the direction of the grain. Also, in hardwoods the direction of the rays or the direction of the large pores definitely indicate the grain. In certain softwoods (those which may contain pitch pockets as mentioned above) the resin ducts are positive indicators of the grain. (These resin ducts appear as fine brownish lines similar to tiny pin scratches.)

If none of these indicators is observed, apply the point of a fountain pen to the wood, and note the direction in which the ink runs in the wood cells. The mark thus produced may be only about $\frac{1}{8}$ in. long, so place a straightedge along this ink mark and with a pencil draw the line several inches long. Another test is to jab the point of a small sharp knife blade into the tangential face at an angle of about 15 deg to the surface, with the knife in a direction crossways to the length of the piece. Then slowly rotate the knife blade with the cutting edge toward the piece of wood. This will lift up a splinter above the surface, and the direction in which the fibers tend to pull out indicates the direction of the grain. A sharp pointed instrument, such as a scriber, pulled loosely in the general direction of the grain along the surface will closely follow the grain (Fig. 17).

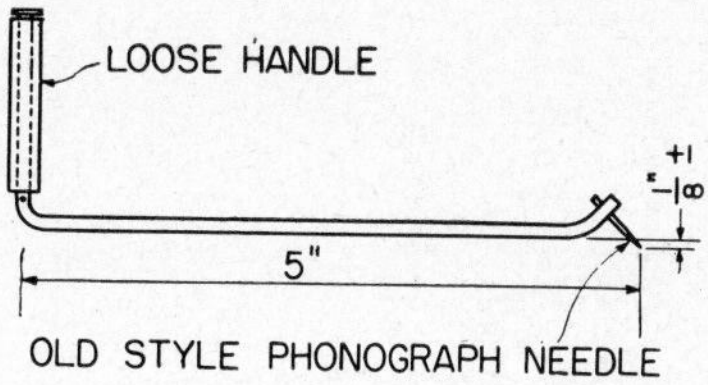

LOOSE HANDLE

5"

OLD STYLE PHONOGRAPH NEEDLE

Fig. 17. Tool for detecting cross grain.

Torn (or chipped) grain on the radial face also indicates spiral grain. Spiral grain should always be measured on the flat-grain surface *farthest* from the pith.

If the annual rings on the cross section run diagonally across the piece, or in other words, if there are no true flat-grain or edge-grain faces, determining the angle of cross grain is more involved.

To measure the angle of *diagonal* grain on such a piece, locate on the end (the cross section of the piece) that corner which is farthest from the pith (see Fig. 18), then from this corner draw a line radially, or at right angle to the rings. On this line, if the size of the piece will permit, measure 1 in. from the corner; then follow the annual ring markings until they intersect that edge farthest from the pith. The ratio $B:A$ indicates the angle of diagonal grain.

To measure the angle of *spiral* grain on a piece of wood whose faces are neither radial nor tangential, mark a point some distance from the end on one of the edges that are neither nearest to nor farthest from the pith and follow along the fibers or a check or parallel to a check to the end of the piece. Then draw a line from the end of this line radially toward the pith. The minimum distance of this radial line from the starting edge D, as shown on Fig. 18, used in conjunction with the distance of the starting point from the end of the stick C will give the angle of spiral grain.

Some wood may contain both diagonal grain and spiral grain. To obtain the true angle of the grain in such a piece, first obtain the angles of the spiral grain and the diagonal grain, both being expressed decimally. Thus, an angle of 1:15 is $\frac{1}{15}$ or 0.066. Square the angle of diagonal grain, and add to it the square of the angle of spiral grain.   Then extract the square root of the sum to obtain the true angle of the grain.

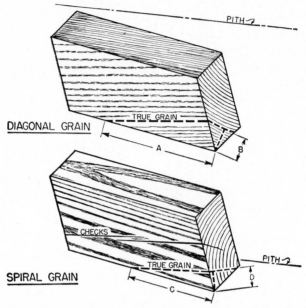

FIG. 18.   Method of measuring the angle of diagonal and spiral grain when the rings are oblique to the edge of the piece.

Table I gives the effect of cross grain on the strength of beams and posts.   It will be noted that even an angle of 1:15 considerably reduces the strength of a timber.

TABLE I

Slope of grain	Strength in % of strength of straight-grain wood*	
	Beams†	Posts and columns‡
1:40	100	100
1:20	85 to 100	100
1:18	80 to 85	up to 100
1:16	76 to 80	up to 100
1:15	74 to 76	87 to 100
1:14	69 to 74	82 to 87
1:12	61 to 69	74 to 82
1:10	53 to 61	66 to 74
1: 8	50 to 53	56 to 66
1: 6	50 to 53	50 to 56

* Cross grain causes a noticeable loss of strength in beams and columns.
† Applies only to middle half of length.
‡ Applies to entire length.

In measuring the angle of cross grain on a timber the average over a distance of a foot or so should be taken, local irregularities in the grain not being measured except in small pieces.

Small pieces of wood are greatly weakened by the presence of dip grain or burls, which are usually the result of defects that may or may not exist in the piece. Figure 19 shows dip grain in one piece resulting from a knot in the adjacent piece cut from the log. The cross grain in the "dip" usually runs at a severe angle and greatly reduces the strength, particularly when this defect is on the tension side of a beam.

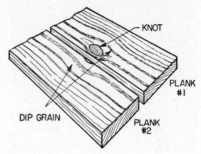

FIG. 19. A knot in one piece of wood may cause dip grain in the adjacent piece.

To determine the strength of a beam of small cross section such as a ladder side rail containing dip grain at the tension face, it should be assumed that the dip cannot take tensile stress, and therefore the depth of the sector should be deducted from the depth of the beam.

In the material for poles, such as flagpoles and telephone poles, spiral grain is acceptable provided the checks or other indicators of the grain do not make one complete turn around the pole in a distance of less than 20 ft.

To ensure straight grain in ladder rungs, they should be hand-shaved, or the piece from which the rung is to be turned should be split (not sawed) in both planes from the board or plank. As ladder rungs are frequently made of red oak, a test for straight grain can be made by wetting one end of the rung with saliva or soapy water and then blowing into the other end. If perfectly straight grain exists, all the wood pores that comprise the circular area at one end of the rung would, if extended, come out within the circular area at the other end of the rung. Thus, when blown into, the air should produce bubbles over the entire area at the other end of the piece. If only part of the area shows bubbles, or if no bubbles show, cross grain is evidenced. (Remember: This can be done only with *red* oak.)

Brashness. This is an abnormal condition which causes wood in bending to break suddenly and completely across the grain when deflected only a relatively small amount. Brashness is usually associated with slow-grown hardwoods, very fast- or very slow-grown softwoods, and wood with pre-existing compression failures. Wood that is exceptionally light weight for its species is usually brash, as is wood that has been exposed to high tempera-

tures for a long time, such as wood ladders used in boiler rooms.

A piece of tough wood and a piece of brash wood may have identical strengths in static bending, but under impact the brash piece will fail at a much lower stress. Under ordinary conditions the strength of wood in tension is from two to four times the compressive strength, but this ratio is greatly reduced in brash wood.

FIG. 20. A typical fracture in tough wood. Note the relatively long splinters.

The length of the splinters produced on the tension side of a beam that has failed in bending is proportional (approximately) to the span of the beam. Thus, in beams of very short span the splinters produced are very short. Dry wood usually breaks more suddenly than moist wood, but still the fracture contains splinters (see Fig. 20).

FIG. 21. A typical fracture in brash wood. Note the absence of any splinters. In making the splinter test with the knife small chips of wood fly out.

On the other hand, brash wood usually fails with a splinterless fracture, which gives a suggestion of crystallization, with the open ends of the cells or pores conspicuous, especially under a magnifying glass (see Fig. 21).

As mentioned previously, summerwood is more dense and has greater strength than springwood. In softwoods, wide rings may be predominately springwood, which is one of the reasons why wide-ring softwoods are brash. On the other hand, the difference between wide- and narrow-ring ring-porous hardwoods is usually mostly in the summerwood, which means that in very narrow rings the wood is mostly porous springwood and consequently low in strength and brashy. See Fig. 22, which shows sections of fast-grown and slow-grown red oak, the former having a much higher percentage of strong summerwood.

Splintering occurs only in the tension half of a fracture, even in tough wood, so the abrupt failure on the compression half should not in itself be taken as an indication of brashness, pre-existant compression failure, or other defect.

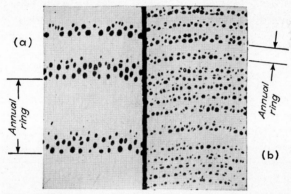

FIG. 22. Cross sections of (a) fast-grown and (b) slow-grown ring-porous hardwood indicate the greater percentage of strong, dense summerwood in the former. (*Forest Products Laboratory.*)

Abrupt failures are not usually caused by exposure to excessive heat unless the temperature has been high enough to darken the wood throughout. Prolonged moderately high temperatures, however, may make the wood brash. Heating wet wood may set up damaging stresses within the piece, especially if it contains the pith and if it is heated to about 160°F.

Another test for the presence of brash wood is to make a knife test as described under "Cross Grain." If the wood is tough, a splinter will be raised up but will not tear out unless the knife blade is rotated too far. On the other hand, if the wood is brash, the piece will fly out with a faint snap, leaving a recess of punky wood (Fig. 21). Of course there are various degrees of brashness; therefore there is no definite line of demarcation between tough and brash wood.

Compression Failures. Next to cross grain and brashness, compression failures are perhaps the most hazardous defects in structural wood. This defect fortunately is not too common, but its danger lies in the fact that it is sometimes extremely difficult, if not actually impossible, to detect such a defect with the unaided eye.

As mentioned previously, most species of wood are from two to four times as strong in tension as in compression. (This is just the opposite to concrete and cast iron.) If a piece of wood, be it a storm-tossed tree, a falling tree, a log being roughly handled, or a plank or ladder being dropped, is subjected to excessive bending stress, the fibers on the compression side of the neutral plane will be compressed to such a degree that they are buckled. But as these wood fibers have no place to be displaced to, it does not actually collapse, and the timber may be subjected to additional load until final failure occurs on the tension side.

If, however, the load is relieved after the compression failure occurred, the piece will be in apparently perfect condition. In fact, even though the average carpenter or rigger is told that a compression failure exists in a piece of finished lumber being examined, the chances are 100 to 1 that he will be unable to locate it. If the piece being inspected is rough-sawed, it is impossible to find the point of failure.

Yet the piece has failed halfway through, and if the beam is reversed so that the compression failure is on the tension side, it may collapse without warning under a very small load. Such a break will appear without splinters and may be confused with a fracture in brash wood.

To inspect a piece of finished lumber for a suspected compression failure, move the wood in relation to a light source, or vice versa, so that there will be reflected glare on the surface of the area being examined. A compression failure can then be detected as a very faint irregular line running crossways to the piece at an angle of 70 to 90 deg to the edge (see Fig. 23). A liberal application of carbon tetrachloride or tung oil to the surface of wood often makes compression failures more easily visible.

Compression failures may occur in several places, their spacing depending upon the length of the beam. Usually they occur only on one face and halfway across the two adjacent faces, assuming, of course, that the injury occurred to the piece in its existing dimensions. If, however, the compression failure occurred during the felling of the tree or the handling of the log and consequently caused damage halfway through the log, then perhaps the beam being inspected has been cut from the defective side of the log. In this case, the compression failure should be visible on all four finished faces of the piece.

Wood containing a compression failure is extremely weak under impact. Therefore, if the piece suspected of containing a compression failure is not too heavy, hold one end of it with the ques-

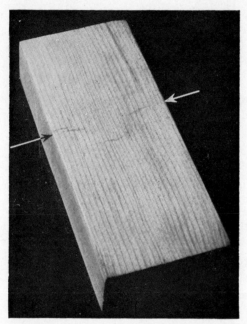

FIG. 23. A compression failure is one of the most difficult defects to detect.

tionable face downward, and allow the other end to drop onto the concrete floor or pavement. The ease with which a piece containing a compression failure is broken may prove a real surprise.

In examining a timber that has failed under load and has an abrupt splinterless fracture, there may be some question as to whether it was caused by a compression failure or by brash wood. To check the condition, examine the width of the annual rings, feel the heft of the piece, and make the knife test for brash wood. If no indication of brashness is observed, it can be assumed that a compression failure existed prior to the final failure. Then carefully examine the wood for several feet each side of the break to locate any secondary compression failures, as described above. Their presence is definite indication that the break was due to a compression failure.

Compression Wood. Compression wood is a defect found only in softwoods. It is recognized by its eccentric annual rings (Fig. 24), which are of a dense nature and are predominately summerwood. Compression wood on the flat-grain and edge-grain faces is dull and lifeless in appearance. It is the one exception to

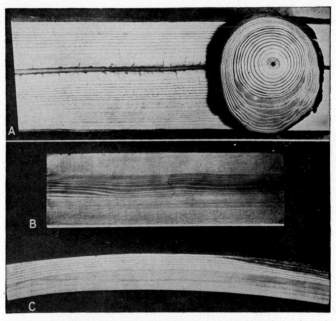

Fig. 24. Compression wood (*A*) Longitudinal and cross sections through part of a tree trunk with compression wood on lower side. (*B*) Cross break in compression wood, and split between compression wood and normal wood due to greater longitudinal shrinkage of compression wood. (*C*) Crook caused by longitudinal shrinkage of the compression wood on the lower side of the piece. (*Forest Products Laboratory.*)

the rule that the strength of any piece of wood, regardless of species, can be judged by its weight. Compression wood is heavy, but is brash and low in strength, particularly under shock.

When a beam containing compression wood fails in bending, the break usually does not extend directly across the piece but either zigzags back and forth producing thick blunt splinters or more often extends directly across the grain from the tension side inward and then branches out in two directions diagonally across the beam like a wide Y. Compression wood, by its very nature, should be excluded from all uses where strength under shock or impact is essential.

The great longitudinal shrinkage of compression wood ($2\frac{1}{2}$ to 20 times normal wood), if located near the center of the piece, may cause this part of the wood to fail in tension, thereby producing cross cracks in the compression wood (Fig. 24*B*).   If located near the edge of a small piece, the compression wood in shrinking will cause crooking or bowing (Fig. 24*C*).   Also, owing to the abnormal shrinkage, spike knots are frequently twisted from their positions and caused to protrude above the face of the material.

Poles containing compression wood are hazardous for linemen to climb, for if cross checks develop at the surface, the shell of the outer wood may peel off and cause the points of their climbers to lose their hold on the pole.

Decay.   Sooner or later nearly all domestic woods when placed in damp locations develop the defect we call decay or rot.   This is *not* an inorganic oxidizing process such as the rusting of steel or the crumbling of stone.   Instead, decay is a disease of the wood just as "TB" or pneumonia, for instance, are diseases of the human body.

To be susceptible to decay, four conditions must be present: (1) The wood must provide food for the fungus.   Unless treated chemically, most woods are able to do this to some extent. (2) There must be a sufficient amount of moisture present.   Wood that contains less than 20 per cent moisture will not decay.   But neither will wood that is submerged in water.   (3) There must be air present, a moist stagnant air being most effective.   (4) There must be warmth, as wood will not decay in extremely cold climates.

We have all seen the fruiting bodies produced by decay fungi on fallen trees, tree stumps, etc., some being in the form of toadstools, shelves, or crusts.   On the underside of these fruiting bodies millions of tiny spores or germs are produced, which, when matured, become free and are blown about by the wind.   Should they be deposited on susceptible wood, they will germinate and form minute fibrous strands (Fig. 25), which, as they grow, puncture the cell walls and not only feed on the contents of the cells but also actually devour the cell walls.   Also, good wood in contact with rotting wood will soon be likewise infected.

The term "dry rot" is frequently used, but this is a misnomer as no wood that is dry can rot.   It is usually referred to as decay where there is no visible evidence of contact with moisture, such as in the case of portable ladders. The moisture had to be present at some time, but not necessarily continuously.Wood that has been painted while in the "green" or unseasoned state may decay under the coating of paint.

In its advanced stage, decay is more readily recognized, but the only remedy then is the removal and replacement of the rotted members. In the incipient stage, it may not be too late to take remedial action, but the symptoms must be first recognized. Among the surface indications of incipient decay are the following:

1. Small bleached or otherwise discolored areas on the surfaced wood.
2. Zigzag zone lines not far from the ends of structural timbers.

Fig. 25. Mycelium of typical brown rot on a piece of oak. Also typical cracking. (*Forest Products Laboratory.*)

3. Oozing out of extractive liquids from the joints between wood structural members.
4. Dark zones in the wood separated by zones of lighter color tissue caused by certain fungi.
5. Persistently moist appearance of freshly cut sections.

Further decay can be arrested by keeping the wood dry. Advanced decay in the interior of wood structural members can be detected by the following evidence:

1. Sagging of structural members.
2. Jabbing a knife blade into the wood and noting the resistance to penetration. Rotted wood is very soft.
3. Drilling small holes into the timber and observing the resistance to drilling, and the color of the chips removed. An "increment borer," designed for this purpose, is a hand auger with a hollow bit which removes a sample like a small dowel

stick which can then be carefully examined. Dark wood, powder or paste is an indication of internal decay.
4. Striking the timber with the round end of a ball-peen hammer and noting if it sounds "dead" and hollow.
5. Loss of resonance of a stick, when struck, against a concrete floor.

Decayed wood is much more combustible than sound wood and therefore presents a greater fire hazard as well as a structural weakness.

It is a poor practice to leave the end grain of any wood exposed, inasmuch as it readily absorbs moisture from the air. Heavily painting or tarring the ends of lumber is advisable. The end grain of oak in contact with steel beams or plates will cause rapid local corrosion of the metal. Contact with galvanized metal will cause local rotting of oak.

A relative humidity in buildings sufficient to bring about condensation is conducive to decay, in some instances the beams being destroyed in two or three years. It should be remembered that the humidity is much higher near cold-water pipes, skylights, uninsulated roofs, etc., so that condensation and decay usually occur first at such locations.

Wood of low decay resistance or that containing a considerable amount of sapwood should not be used in basement or first-floor construction close to the ground unless chemically treated.

No wood that shows evidence of incipient decay should be used where shock or impact loads are possible.

Ordinary paints, varnishes, and similar protective coatings cannot be relied upon to preserve wood against decay, because they contain no substances poisonous to fungi and they may themselves support the growth of fungi in the presence of dampness. Although protective coatings retard the absorption of moisture by wood, they also retard the drying out of wood that has taken up moisture through an uncoated surface or joint.

For example, joints between end-grain and side-grain wood can rarely be kept tight by coatings; rain water can readily enter the joint and be absorbed by the wood, after which its drying out again may be retarded by a coating of paint on the surfaces exposed to view. In such case painting may hasten rather than retard decay.

Sap Stains and Molds. Generally, stains of the sapwood are caused by fungi similar to those which cause decay, except that they attack mainly the contents of the wood cells, not the cell walls. Consequently sap stains are not considered a defect in most structural wood. Heavily stained wood is likely to be low in shock resistance, however, and should be utilized accordingly. Moreover, it may be accompanied by incipient decay inasmuch as the same factors favor the development of stain and decay fungi. But where wood is to be varnished, shellacked, or stained to improve its appearance, such as the trim in house construction, sap stain is, of course, considered a serious defect.

Sap stains are found in various colors: blue, brown, yellow, pink, red, etc., but blue (which includes shades almost black) is by far the most common. Stains penetrate deep into the wood and cannot be surfaced off.

Molds may be yellow, pink, purple, green, or black and can usually be recognized on the sapwood by the presence of cottony or powdery surface growths. These are also caused by fungi. Molds usually appear only on the surface of the wood.

Staining of wood may occur very rapidly, a carload of unseasoned lumber being infected while in transit.

In using stained wood for structural purposes, keep in mind that in order for the stain to propagate the wood must have been exposed to conditions which are favorable to decay. Therefore, thoroughly examine the wood for evidence of incipient decay.

Termite and Other Parasite Attack. The termite is perhaps the most troublesome insect we have to contend with in structural timbers. There are two general types of termites, the subterranean and the nonsubterranean. Only the former is found in the northern part of the country.

These insects begin their attack of the wood from the earth, either directly or by building shelter tubes of particles of earth, pieces of leaves, etc., over the intervening masonry in order to reach the wood.

Figure 26 shows the two most common forms of termites. The winged adult is probably the most frequently observed, while the grayish-white worker is the most destructive. Termites are soft-bodied antlike creatures, which conceal themselves within the wood, in the earth, or in the shelter tubes. The workers are blind, shun the light, and are seldom seen except when a structure or building is demolished or altered and the timber in which they are living is suddenly cut into or the soil excavated.

The winged adult form is brownish or blackish in color with

elongated body and with long white wings extending beyond it at the rear. At certain seasons, usually spring and fall, the winged sexual adults migrate in large numbers and at such times may be observed for a short period of several hours. They then lose their wings, enter crevices between timbers, and breed new colonies. Termites frequently eat away the entire inside of a timber, leaving nothing but a thin shell of wood (and possibly paint) and with no exterior evidence of attack until it ultimately fails structurally.

Observing the migrating termites, the wings that the termites have shed just before reentering the wood; the pellets of fine, digested, excreted wood similar to sawdust at or on the floor below

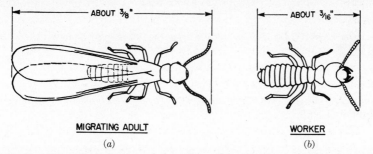

MIGRATING ADULT

(a)

WORKER

(b)

Fig. 26. Two principal forms of the subterranean termite. (a) Winged sexual adult commonly observed when migrating. (b) The tiny but destructive worker, which never comes out into the light and hence is seldom seen.

joints in the wood; holes in the surface of the wood about the size of BB shot; or the sagging or collapse of structural members are evidence of termite attack. Also, the earthlike shelter tubes bypassing masonry to provide communication between the ground and the wood are unmistakable signs of infestation.

Combating termites is nearly a hopeless task unless all the infested wood is removed and burned. When contact with a moisture supply from the earth, leaking water pipe, or roof leak is cut off, the termites depending upon such moisture supply will die. Among the poisons used in combating termites are orthodichlorobenzene, Paris green, sodium fluosilicate, and carbontetrachloride. Often termite damage is confused with decay, which is quite different and is caused by fungi.

Termite exterminators usually excavate to a considerable depth completely around the building in order to apply their chemicals properly. This is not only expensive but also may cause destruction of costly shrubs, trees, lawns, etc. Assuming that the termites obtain their moisture from the ground, and not from a leaking roof, leaking water pipe, or similar source, then it might be desirable to consider jacking up the house $\frac{1}{8}$ in. or so off the foundation and inserting a sheet-metal termite shield between the sill and the foundation, after which the house is again lowered onto its foundation. Such a termite shield should extend about 1 in. beyond the inside and outside faces of the foundation, and these projecting edges bent downward at a 45-deg angle to shed rain water. Keep in mind that insects cannot pass from one side of a thin sheet to the other side via the knifelike edge. The termite-shield method of combating termites should appeal to the rigger.

Warping. As mentioned previously, when wood dries it shrinks more tangentially than radially and an almost insignificant amount longitudinally. Unless influenced by outside factors, dimensional changes caused by shrinkage follow a definite pattern (see Fig. 27).

Edge-grain planks shrink more on the thickness.

Flat-grain planks shrink more on the width.

Boxed heart timbers cause all faces to become slightly convex, except if large checks develop the faces will become somewhat concave.

Boxed heart planks cause the wide faces to become convex.

Timbers with their rings running diagonally become diamond shape.

Flat-grain boards "cup" so that the rings tend to straighten out (with the convex side toward the heart or pith of the tree).

Long sticks having compression wood near one face bow or crook toward that face (concave on that face).

Boards or timbers having spiral grain twist in a direction so as to increase the angle of checks and other evidence of the direction of the grain.

IDENTIFICATION OF WOOD

Most riggers can readily identify spruce, yellow pine, Douglas fir, and a few other species of wood, while the average carpenter has a much broader knowledge of the various species. The identification is more or less by general appearance rather than by any technical knowledge. In fact, we have all heard carpenters say that they can distinguish between longleaf and shortleaf yellow pine. Perhaps, but the experts at the Forest Products Laboratory say that it is impossible to differentiate between these species unless the pith can be observed and even then it requires a chart to make the determination.

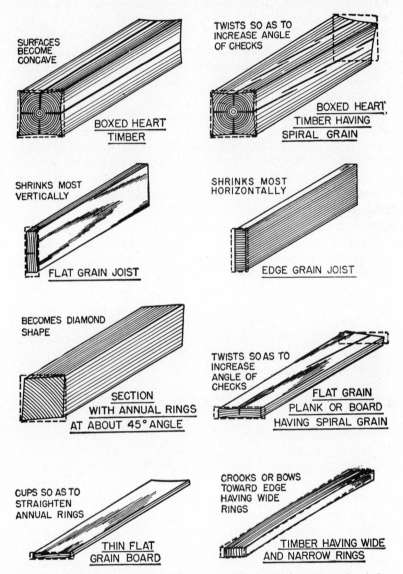

SURFACES BECOME CONCAVE

BOXED HEART TIMBER

TWISTS SO AS TO INCREASE ANGLE OF CHECKS

BOXED HEART TIMBER HAVING SPIRAL GRAIN

SHRINKS MOST VERTICALLY

FLAT GRAIN JOIST

SHRINKS MOST HORIZONTALLY

EDGE GRAIN JOIST

BECOMES DIAMOND SHAPE

SECTION WITH ANNUAL RINGS AT ABOUT 45° ANGLE

TWISTS SO AS TO INCREASE ANGLE OF CHECKS

FLAT GRAIN PLANK OR BOARD HAVING SPIRAL GRAIN

CUPS SO AS TO STRAIGHTEN ANNUAL RINGS

THIN FLAT GRAIN BOARD

CROOKS OR BOWS TOWARD EDGE HAVING WIDE RINGS

TIMBER HAVING WIDE AND NARROW RINGS

Fig. 27. In drying, wood warps in a certain predetermined manner unless influenced by outside factors.

It is possible to identify most species of wood by observing the color (particularly of the heartwood), weight, odor, presence or absence of pores, and arrangement of pores, rays, resin ducts, and other factors.   To make an identification of an unknown species, either of the two keys on the following pages may be used.   The first key (Table II) is for use with the naked eye and is more difficult to use.   The second key (Table III) is for use when a magnifying glass is available.

To use either key, first look under I and II and determine whether the wood in question is "Wood with Pores" or Wood without Pores."   If the latter, then look under A and B to decide if "Resin Ducts Present" or "Resin Ducts Normally Absent" (Table III).   If the former, then under 1 and 2 decide if the resin ducts are numerous or not numerous.   If they are numerous then refer to AA and BB to determine if they are hard pine or soft pine. Continue this procedure until the species of wood is determined.

As an example, let us consider a piece of wood removed from an old building being demolished.   By means of a plane or sharp knife, cut and clean the surface on the two faces and end of the timber so that its structure can be carefully examined.   No magnifier is at hand, so the first key (Table II) will be used.

. Under I or II, are pores visible on the end of the piece?   No, so we enter the key under II.

Under A, B, or C, are the rays visible or conspicuous?   The answer is no, so we proceed to C.

Under 1 or 2, are the annual rings clearly defined?   Yes, so we proceed to 2.

Now, under AA and BB, is the heartwood distinctly darker than the sapwood?   (If the piece being examined is small, it may be difficult to answer this question, so proceed under both AA and BB by trial-and-error method.)   Yes, the heartwood is darker, so advance to AA.

Under (a) and (b), is the wood pitchy?   Yes, so proceed to (a).

Under (aa), (bb), and (cc) check the color of the wood.   The heartwood is reddish brown so we proceed to (aa).

Under (a3) and (b3) the summerwood is found to be very conspicuous, so we advance to (b3), knowing that our wood is a hard pine.

Under (b3), consider the weight of the wood (a5) or (b5).   It is heavy, so continue under (b5).   This indicates one of the eastern species of pine (including southern pine).

To distinguish longleaf pine from shortleaf and loblolly pine (North Carolina pine is loblolly pine), it is necessary first to measure the over-all diameter of the pith and the diameter of the second annual ring. This measurement is taken at the outside of the second year's band of summerwood (see Fig. 28). Then using the chart, draw a horizontal line indicating the pith diameter; also draw a vertical line indicating the diameter of the second annual ring. If these lines intersect above the diagonal line, the wood is longleaf pine; if below the diagonal line, it is either shortleaf or loblolly pine.

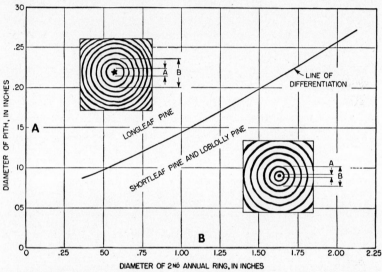

FIG. 28. Identification of longleaf and shortleaf pine when pith and second annual ring can be measured.

To identify the individual species of spruce, the following facts may be of help:

Eastern and Engelmann Spruce. Heartwood is nearly the same color as the sapwood, but usually not clearly defined. Resin ducts not numerous, scattered singly or in tangential groups of 2 to 20 (as observed on the cross section), but not visible without a magnifier. Also, they may appear as whitish specks in the summerwood. Split tangential surface is not dimpled.

Sitka Spruce. Heartwood is pale reddish color, slightly darker than sapwood. The split, tangential surface, especially through the summerwood of narrow rings, is characteristically indented or dimpled. Resin ducts are rather inconspicuous. Resinous odor and taste. Silky sheen on split surfaces.

Table II. Key for the Identification of Woods without the
Aid of a Hand Lens*

(*Forest Products Laboratory*)

HARDWOODS

1. Pores visible.
A. Ring-porous; that is, the pores at the beginning of each annual ring are comparatively large, forming a distinct porous ring, and decrease in size more or less *abruptly* toward the summerwood. (This feature is often more distinct in the outer sapwood where the pores are more open.)
  1. Summerwood figured with wavy or branched *radial* bands.
    AA. Many rays broad and conspicuous. Wood heavy to very heavy...The OAKS
      (a) Wood without reddish tinge. The large pores mostly closed up (exception, chestnut oak).........................The WHITE–OAK GROUP
      (b) Wood with reddish tinge, especially near knots. The large pores mostly open (exception, black jack oak).............The RED–OAK GROUP
    BB. Rays not noticeable. Color grayish brown. Wood moderately light.
                                                                  CHESTNUT
  2. Summerwood figured with short or long wavy *tangential* lines or bands, in some woods more pronounced toward the outer part of the annual ring.
    AA. The heartwood not distinctly darker than the sapwood (the sapwood may be darker than the heartwood on account of sap stain). The wavy tangential bands conspicuous throughout the summerwood. Color yellowish or greenish-gray. Wood moderately heavy.........................HACKBERRY
                                                                  SUGARBERRY
    BB. The heartwood distinctly darker than the sapwood.
      (a) Wood with spicy odor and taste: moderately heavy. Heartwood silvery brown...SASSAFRAS
      (b) Wood without spicy odor or taste.
        (aa) Heartwood bright cherry red to reddish brown. Pores in springwood all open and very distinct. Sapwood narrow. Wood very heavy.
          (a3) Pith large, usually over 0.2 and often about 0.3 in. in diameter.
                                                                  COFFEETREE
          (b3) Pith small, usually under 0.15 and often less than 0.1 in. in diameter.............................HONEY LOCUST
        (bb) Heartwood russet to golden brown. Pores entirely closed up except in outer sapwood. Sapwood very narrow.
          (a3) Wood from very heavy to very, very heavy and exceedingly hard. Tangential bands confined to, or more pronounced in, the outer portion of the annual ring. Rays barely distinct.
            (a4) Heartwood golden brown with reddish brown streaks; yellowish color imparted in a few minutes to a wet rag or blotter.............................OSAGE ORANGE
            (b4) Heartwood russet brown without reddish brown streaks; color not readily imparted to a wet rag or blotter.
                                                                  BLACK LOCUST
          (b3) Wood lighter but still classed as heavy and hard. Tangential bands uniformly distributed throughout the summerwood. Rays very distinct......................RED MULBERRY
        (cc) Heartwood grayish brown. Tangential bands short and confined mostly to the outer portion of the summerwood (inconspicuous in black ash).
          (a3) Sapwood narrow, rarely over $\frac{3}{4}$ in. wide.
            (a4) Annual rings mostly wide, especially within the first few inches from the center. Pores containing glistening tyloses. Pith usually three-sided. Wood moderately light.
                                                                  HARDY CATALPA

*Unless it is otherwise directed, all observations as to structure should be made on the end surface of rings of average width, cut smoothly with a very sharp knife; and all observations as to color should be made on a freshly cut longitudinal surface of the heartwood.

(b4) Annual rings mostly narrow, even near the center. Pores partly filled with tyloses, not glistening. Pith usually round. Wood moderately heavy..........BLACK ASH

(b3) Sapwood over an inch, usually several inches wide. Wood heavy and hard...................................WHITE ASH
GREEN ASH

(dd) Heartwood brown with reddish tinge. Tangential bands long and very conspicuous throughout the summerwood.

(a3) Sapwood very narrow. The porous ring of springwood from 2 to 4 pores wide. Inner bark slimy when chewed. Wood moderately heavy...........................SLIPPERY ELM

(b3) Sapwood moderately wide, the porous ring of springwood only one pore wide except in very wide rings.

(a4) Wood heavy. Pores in springwood inconspicuous because comparatively small, not close together, and plugged with tyloses.................................CORK ELM

(b4) Wood moderately heavy. Pores in springwood fairly conspicuous because larger than in rock elm, close together and open.............................WHITE ELM

3. Summerwood not figured with radial or tangential bands distinctly visible without a lens (fine tangential lines may be seen in hickory and persimmon with a hand lens). See figure 5.

AA. Sapwood wide, over 2 in.

(a) Heartwood black, or brownish black (usually very small). *Tangential surface* marked with fine bands which run across the grain and are due to the storied arrangement of the rays. Wood very, very heavy..PERSIMMON

(b) Heartwood reddish brown. *Tangential surface* not marked with fine cross bands. Wood very heavy.........................The HICKORIES

BB. Sapwood narrow, rarely over $\frac{3}{4}$ in. wide. Heartwood grayish brown. Wood moderately heavy.......................................BLACK ASH

B. Diffuse-porous; that is, no ring of large pores is formed at the beginning of each annual ring.

1. Individual pores plainly visible.

AA. *Tangential surface* marked with fine bands which run across the grain and are due to the storied arrangement of the rays. Heartwood black, or brownish black (usually very small). Sapwood wide. Wood very, very heavy and hard.
PERSIMMON

BB. *Tangential surface* not marked with fine cross bands.

(a) Heartwood reddish brown. Sapwood wide. Wood heavy.
WATER HICKORY

(b) Heartwood chocolate brown. Sapwood from moderate in width to narrow. Wood heavy and hard..........................BLACK WALNUT

(c) Heartwood light chestnut brown. Sapwood narrow. Wood moderately light and soft...BUTTERNUT

2. Individual pores barely visible under conditions of good light and a very smoothly cut end surface.

AA. Pores not crowded. Heartwood reddish brown. Wood heavy.

(a) Inner bark with wintergreen flavor. Pith flecks very rare.
YELLOW BIRCH
SWEET BIRCH

(b) Inner bark without wintergreen flavor. Pith flecks usually abundant.
RIVER BIRCH

BB. Pores crowded. Heartwood grayish. Wood light.........COTTONWOOD

II. Pores not visible.

A. Rays comparatively broad and conspicuous. Color in various shades of light reddish brown.

1. The rays crowded. No denser and darker band of summerwood noticeable. Wood usually lock-grained: moderately heavy.........................SYCAMORE

2. The rays not crowded. A distinct, denser, and darker band of summerwood present. Wood usually fairly straight-grained: heavy...........................BEECH

B. Rays not conspicuous but distinctly visible.
    1. Heartwood deep, rich, reddish brown. Sapwood narrow, usually less than 1 in. wide. Annual rings clearly defined. Rays very distinct. Wood moderately heavy.
BLACK CHERRY

    2. Heartwood dingy, reddish brown, often with darker streaks. Sapwood moderately wide, usually over 1 in. Annual rings not clearly defined. Rays relatively not very distinct. Wood moderately heavy..............................RED GUM

    3. Heartwood light reddish brown. Sapwood wide. Annual rings clearly defined by a thin, darker reddish-brown layer. Rays very distinct.
        AA. Wood heavy and hard; difficult to cut across the grain. Pith flecks very rare.
SUGAR MAPLE

        BB. Wood lighter and softer, rather easy to cut across the grain. Pith flecks often abundant...SILVER MAPLE
RED MAPLE

    4. Heartwood light yellowish brown with greenish tinge. Sapwood usually over 1 in. wide. Annual rings clearly defined. Rays fairly distinct. Wood moderately light.
YELLOW POPLAR

    5. Heartwood creamy brown with occasional darker streaks. Sapwood wide and not sharply defined from the heartwood. Rays fairly distinct. Wood light.
BASSWOOD

C. Rays not distinctly visible.
    1. Annual rings not clearly divided into a band of soft springwood and denser and darker band of summerwood and, therefore, not conspicuous.
    AA. The heartwood distinctly darker than the sapwood.
        (a) Heartwood reddish brown. Wood straight-grained; heavy.
            (aa) Inner bark with wintergreen flavor. Pith flecks rare.
YELLOW BIRCH
SWEET BIRCH

            (bb) Inner bark without wintergreen flavor. Pith flecks usually abundant.
RIVER BIRCH

        (b) Heartwood grayish brown.
            (aa) Wood cross-grained; moderately heavy.BLACK GUM
COTTON GUM (TUPELO)

            (bb) Wood fairly straight-grained; light..............COTTONWOOD

    AA. The heartwood not distinctly darker than the sapwood.
        (a) Wood odorless and tasteless; light and soft. Color yellowish.
YELLOW BUCKEYE
OHIO BUCKEYE

CONIFERS

        (b) Wood with spicy odor and taste; moderately light in weight. Color pale brown.................................PORT ORFORD CEDAR

        (c) Wood with resinous odor; heavy to very heavy. Color creamy brown.
PIÑON (PINE)

    2. Annual rings clearly divided into a band of soft springwood and a denser and darker band of summerwood. Although the summerwood may not be pronounced, yet the annual rings are always clearly defined by it.
    AA. The heartwood distinctly darker than the sapwood.
        (a) Wood "pitchy," as indicated by the resinous odor and by exudations of resin at the ends, especially from the sapwood, although on cuts made after the wood is seasoned the resin does not come out unless the wood is heated.
        (aa) Heartwood creamy or orange-brown to reddish brown. Resin ducts abundant, visible as minute openings or, more often, as darker or lighter colored specks, or as brownish lines on longitudinal surfaces. Sapwood widely variable in width....................The PINES
            (a3) The summerwood inconspicuous and not much harder than the springwood...........................*The SOFT PINES:*
            (a4) Wood soft and moderately light; straight-grained. Annual rings of moderate width. Heartwood light reddish brown.,.......,...........WESTERN WHITE PINE
LIMBER PINE

(b4) Wood hard and moderately heavy to very heavy, often cross-grained. Annual rings narrow.

    (a5) Heartwood reddish brown. Tangential surface has numerous slight depressions which give it a dimpled appearance especially noticeable on split surfaces.

                         BRISTLE–CONE PINE

    (b5) Heartwood creamy brown. Tangential surface not dimpled. Wood sometimes very heavy.

                         PIÑON (PINE)

(b3) The summerwood conspicuously darker and harder than the springwood. (This feature is not so noticeable in the sapwood of old trees as in the heartwood.)

                         The HARD PINES:

    (a5) Wood moderately light.   *WESTERN SPECIES:*

        (a6) The sapwood usually less than 2 in. wide (mostly about 1 in.). Tangential surface has numerous slight depressions, which give it a dimpled appearance, especially noticeable on split surfaces. Openings of resin ducts not visible without a lens....................LODGEPOLE PINE

        (b6) The sapwood usually over $2\frac{1}{2}$ in. wide (mostly over 3 in.). Tangential surfaces rarely dimpled. Openings of resin ducts often visible without a lens............WESTERN YELLOW PINE

    (b5) Wood moderately heavy to very heavy. Heartwood orange brown to reddish brown.

                     EASTERN SPECIES:
                     NORWAY PINE
                     PITCH PINE
                     SHORTLEAF PINE
                     LOBLOLLY PINE
                     POND PINE
                     LONGLEAF PINE
                     SLASH PINE

       (For distinguishing longleaf from loblolly and shortleaf pine, see Fig. 27.)

(bb) Heartwood orange-reddish to red. Resin ducts not abundant, occasionally visible as whitish specks in the summerwood. Sapwood usually over 1 in. wide. Wood moderately heavy...DOUGLAS FIR

(cc) Heartwood russet brown. Resin ducts not abundant; usually not visible without a lens. Sapwood usually less than 1 in. wide. Wood moderately heavy.

    (a3) Annual rings narrow..................WESTERN LARCH

    (b3) Annual rings moderately wide.................TAMARACK

(b) Wood not "pitchy" or resinous, although resin may exude from the bark.

    (aa) Heartwood deep reddish brown; without characteristic odor or taste. Annual rings regular in width and outline. Sapwood over 1 in. wide. Wood moderately light............................REDWOOD

    (bb) Heartwood light brown to dark, dingy brown, with or without reddish tinge. Odor characteristic but not resinous or "pitchy."

        (a3) Odor somewhat rancid; heartwood tasteless. Annual rings mostly irregular in width and outline. Sapwood usually over 1 in. wide. Color highly variable from pale brown with or without reddish tinge to blackish brown. Weight variable from moderately light to heavy. Longitudinal surfaces feel and appear waxy.......................BALD CYPRESS

        (b3) Odor aromatic (like cedar shingles); heartwood slightly bitter in taste. Annual rings narrow but regular in width. Sapwood rarely over 1 in. wide. Wood very light in weight. Longitudinal surfaces not appearing waxy.

           (a4) Heartwood brown with reddish tinge.

                     WESTERN RED CEDAR

   (b4) Heartwood brown, rarely with reddish tinge.

      ARBORVITAE (NORTHERN WHITE CEDAR)

 BB. The heartwood not distinctly darker than the sapwood.

   (a) Wood resinous, as indicated by the odor or exudations of resin at the ends, especially from the sapwood.

    (aa) Tangential surface has numerous slight depressions, which give it a dimpled appearance, especially noticeable on split surfaces. The heartwood sometimes slightly darker than the sapwood. Wood moderately heavy........................LODGEPOLE PINE

    (bb) Tangential surface not dimpled.

     (a3) Color pale brown, almost white. Annual rings mostly moderately wide. Wood light in weight...ENGELMANN SPRUCE

     (b3) Color creamy brown. Annual rings mostly narrow. Wood heavy to very heavy.......................PIÑON (PINE)

   (b) Wood not resinous.

    (aa) Odor and taste spicy. Color pale brown. Wood moderately light.

      PORT ORFORD CEDAR

    (bb) Odor and taste not spicy, although a characteristic odor may be noticeable.

     (a3) Wood whitish, at least in the springwood; the summerwood may be dark reddish brown, especially in pieces of rapid growth, in which case there is a decided contrast between springwood and summerwood.

      (a4) Freshly cut surface of dry wood has a mild, rank odor. Little contrast between springwood and summerwood. Growth rings of moderate width. Wood light.

         ALPINE FIR

      (b4) Freshly cut surface of dry wood does not have a rank odor; decided contrast between springwood and summerwood. Growth rings fairly wide. Wood moderately light.

       (a5) Outer bark contains whitish layers.....WHITE FIR

       (b5) Outer bark contains thin, very dark reddish-brown layers............................GRAND FIR

     (b3) Wood has reddish hue, the springwood as well as the summerwood being colored. Moderately light to moderately heavy. Fresh pieces have a sour odor.

      (a4) Wood coarse and splintery, often cup-shaken.

        EASTERN HEMLOCK

      (b4) Wood not very coarse or splintery, usually not cup-shaken.

        WESTERN HEMLOCK

TABLE III.　KEY FOR THE IDENTIFICATION OF WOODS WITH THE AID OF A HAND LENS*

(*Forest Products Laboratory*)

HARDWOODS

 I. Wood with pores. The pores are conspicuously larger than the surrounding cells, although in some species they are not visible without magnification. Neither the pores nor other cells are in continuous radial rows.

  A. Ring-porous; that is, the pores at the beginning of each annual ring are comparatively large, forming a distinct porous ring, and decrease in size more or less *abruptly* toward the summerwood.

   1. Summerwood figured with wavy or branched *radial* bands. The bands visible without a lens on a smoothly cut surface.

 *Unless it is otherwise directed, all observations as to structure should be made on the end surface of rings of average width cut smoothly with a very sharp knife and all observations as to color should be made on a freshly cut longitudinal surface of the heartwood.

AA. Many rays very broad and conspicuous. Wood heavy to very heavy.

The OAKS

(a) Pores in the summerwood very small and so numerous as to be exceedingly difficult to count under a lens; pores in the springwood usually densely plugged with tyloses. Heartwood brown without reddish tinge.

The WHITE–OAK GROUP

(b) Pores in the summerwood larger, distinctly visible with (sometimes without) a hand lens and not so numerous but that they can readily be counted under a lens; pores in springwood mostly open, tyloses not abundant. Heartwood brown, with reddish tinge especially in vicinity of knots.

The RED–OAK GROUP

BB. All rays very fine and inconspicuous. Color grayish brown. Wood moderately light...CHESTNUT

2. Summerwood figured with long or short wavy *tangential* bands which include the pores. The bands visible without a lens on a smoothly cut end surface.

AA. Careful examination with a hand lens shows the pores of the summerwood to be joined in more or less continuous bands, and the bands to be evenly distributed throughout the summerwood.

(a) Sapwood moderate in width or narrow, mostly less than 3 in.; heartwood distinct, light to deep reddish brown. Rays not distinct without a lens.

The ELMS:

(aa) Large pores in the springwood usually in one row except in very wide rings.

(a3) Rows of pores in the springwood conspicuous because the pores are large enough to be plainly visible without a lens; they are mostly open, containing only a few tyloses; and they are fairly close together. Sapwood from 1 to 3 in. wide. Wood moderately heavy; fairly easy to cut...............WHITE ELM

(b3) Rows of pores in the springwood inconspicuous because the pores are small, being barely visible without a lens; they are mostly closed with tyloses, especially in the heartwood; and they are often somewhat separated. Sapwood from $\frac{3}{4}$ to $1\frac{1}{2}$ in. wide. Wood heavy and difficult to cut................CORK ELM

(bb) Large pores in springwood in several rows; mostly open, containing few tyloses. Sapwood usually less than 1 in. wide, often only $\frac{1}{2}$ in. wide. Wood moderately heavy. Inner bark mucilaginous when chewed....................................SLIPPERY ELM

(b) Sapwood wide, over 3 in., heartwood indistinct, yellowish or greenish gray. Pores in springwood mostly open, in several rows except in occasional narrow rings where they may form only one row. Rays distinct without a lens. Wood moderately heavy.......................HACKBERRY

SUGARBERRY

BB. Careful examination with a hand lens shows the pores of the summerwood to be joined in more or less interrupted bands or in rounded groups of from 3 to 20 (especially in mulberry and coffeetree), the groups so arranged as to form tangential bands. In either case the bands are more pronounced in the outer portion of the summerwood than in the middle of the annual ring, where the pores are often isolated or in rounded groups.

(a) Large pores in the springwood containing numerous tyloses. Sapwood narrow, usually less than 1 in. wide.

(aa) Wood very, very heavy and exceedingly hard to cut across the grain. Rays not very distinct without a lens.

(a3) Heartwood golden brown with reddish brown streaks; coloring matter readily soluble in cold water........OSAGE ORANGE

(b3) Heartwood russet brown; coloring matter not readily soluble in cold water..............................BLACK LOCUST

(bb) Wood heavy, but lighter than the above and fairly easy to cut across the grain. Color russet brown. Rays very distinct without a lens.

RED MULBERRY

(cc) Wood moderately light and easy to cut across the grain. Color grayish brown. Rays not distinct without a lens.... HARDY CATALPA

(b) Large pores in the springwood open, containing no tyloses but occasionally a bright-red gum. Heartwood cherry-red to reddish brown. Wood very heavy.

(aa) Pores in the outer portion of the summerwood mostly joined into bands, the individual pores of which are not distinctly visible with a lens magnifying 15 diameters. Rays mostly very distinct. Pith small, usually under 0.15 in. Sapwood from ¾ to 2 in. wide on ties.
HONEY LOCUST

(bb) Pores in the outer portion of the summerwood only occasionally joined into bands, the individual pores being distinctly visible with an ordinary hand lens. Rays of uniform width, inconspicuous. Pith large, usually over 0.2 in. Sapwood from ½ to 1 in. wide on ties.
COFFEETREE

CC. Careful examination with a hand lens shows the pores of the summerwood to be isolated or in radial rows of 2 or 3, but surrounded by parenchyma in such a manner as to appear in wavy tangential bands usually more distinct without a lens than with a lens.

(a) Parenchyma projecting tangentially from the pores in comparatively long lines often joining pores widely separated. Sapwood several inches wide; heartwood grayish brown, occasionally with reddish tinge. Wood heavy and hard...WHITE ASH
GREEN ASH

(b) Parenchyma not projecting tangentially from the pores or only slightly so. Sapwood less than 1 in. wide; heartwood silvery brown. Wood moderately heavy.

(aa) Rays fine but distinct without a lens; wood has a spicy odor and taste.
SASSAFRAS

(bb) Rays not visible without a lens; wood does not have a spicy odor and taste...BLACK ASH

3. Summerwood figured with numerous fine, light-colored tangential lines (parenchyma), which do not embrace the pores. Pores in the summerwood not much smaller than those in the springwood, usually visible without a lens. (Water hickory and persimmon are also classed as diffuse-porous woods.) Wood very heavy to very, very heavy.

AA. Lines of parenchyma inconspicuous even under a lens. Rays in tiers, appearing on tangential surface as fine bands running across the grain. Heartwood black or brownish black.......................................PERSIMMON

BB. Lines of parenchyma conspicuous under a lens, barely visible without a lens. Rays on tangential surface not in tiers; heartwood reddish brown.
The HICKORIES

4. Summerwood not figured with radial or tangential bands. Pores in summerwood very small, not visible without a lens, isolated, or in radial rows of two or three. Sapwood very narrow, heartwood silvery or grayish brown. Wood moderately heavy...BLACK ASH

B. Diffuse-porous: that is, the pores are of about uniform size and evenly distributed throughout the annual ring, or if they are slightly larger and more numerous in the springwood, they *gradually* decrease in size and number toward the outer edge of the ring.

1. Rays comparatively broad and conspicuous, the widest ones fully two times as wide as the largest pores, appearing on the radial surface as distinct "flakes" or "silver grain" similar to quartered oak, but finer. Color in various shades of light reddish brown.

AA. Practically all rays broad. Pores crowded, decreasing little, if any, in size at extreme outer edge of the annual ring. Wood usually lock-grained, moderately heavy...SYCAMORE

BB. Only part of the rays broad, the others narrower than the largest pores. Pores crowded in the springwood, decreasing in size and number toward the outer edge of the annual ring, thereby giving rise to a harder and darker band of summerwood. Wood usually fairly straight-grained; heavy...............BEECH

2. Rays narrower, but very distinct without a lens, the widest ones of about the same width as the largest pores.

AA. Color light brown with reddish tinge. Springwood and summerwood of uniform density. Sapwood wide.

(a) Wood heavy, difficult to cut across the grain. Only part of the rays broad, the others very fine, scarcely visible with a lens. Pith flecks rarely present.
SUGAR MAPLE

(b) Wood moderately heavy, fairly easy to cut across the grain. Practically all the rays broad but not so broad as in sugar maple, therefore not so prominent but giving the appearance of being more numerous. Pith flecks common.......................................SILVER MAPLE
RED MAPLE

BB. Color deep reddish brown. Springwood slightly more porous than summerwood. Sapwood narrow. Pith flecks common. Wood moderately heavy.
BLACK CHERRY

3. Rays comparatively fine, narrower than the largest pores.

AA. Pores visible without a lens.

(a) Pores comparatively large and conspicuous without a lens, decreasing in size toward the outer limit of each annual ring; not crowded. Fine tangential lines of parenchyma often visible between the pores.

(aa) Sapwood wide, usually over 3 in. in ties. (Pores often in a more or less well-defined zone in the springwood, therefore also classed as ring-porous woods.)

(a3) Heartwood black or brownish black. Rays in tiers, appearing on the tangential surface as fine bands running across the grain. Wood very, very heavy......................PERSIMMON

(b3) Heartwood reddish brown. Rays not in tiers. Wood heavy.
WATER HICKORY

(bb) Sapwood narrow, mostly under 2 in. in ties; white or discolored; heartwood brown.

(a3) Wood heavy and hard. Heartwood chocolate brown.
BLACK WALNUT

(b3) Wood moderately light and soft. Heartwood light chestnut brown......................................BUTTERNUT

(b) Pores smaller, but on careful examination still clearly visible without a lens, at least in the springwood.

(aa) Pores not crowded, decreasing little, if any, in size toward the outer limit of the annual ring. Rays distinct under a lens. Heartwood pale to moderately deep reddish brown. Wood heavy.

(a3) Pith flecks rare. Inner bark has a wintergreen flavor.
YELLOW BIRCH
SWEET BIRCH

(b3) Pith flecks abundant. Inner bark does not have a wintergreen flavor....................................RIVER BIRCH

(bb) Pores crowded, decreasing somewhat in size and number toward the outer limit of each annual ring. Rays very fine, barely visible with a lens. Pith flecks occasionally present but not abundant. Wood light and soft. Color white to light grayish brown.COTTONWOOD

BB. Pores not visible without a lens.

(a) Pores appearing comparatively large and conspicuous under a lens.

(aa) Pores not crowded, decreasing little, if any, in size toward the outer limit of the annual ring. Rays distinct under a lens. Heartwood pale to moderately deep reddish brown. Wood heavy.

(a3) Pith flecks rare. Inner bark has a wintergreen flavor.
YELLOW BIRCH
SWEET BIRCH

(b3) Pith flecks abundant. Inner bark does not have a wintergreen flavor....................................RIVER BIRCH

(bb) Pores crowded, decreasing somewhat in size and number toward the outer limit of each annual ring. Rays very fine, barely visible with a lens. Pith flecks occasionally present but not abundant. Wood light and soft. Color white to light grayish brown.COTTONWOOD

(b) Pores appearing comparatively small under a lens.
  (aa) Heartwood pale reddish brown. Rays very distinct without a lens. Pores not very crowded. Wood moderately heavy.SILVER MAPLE
                                                        RED MAPLE
  (bb) Heartwood dingy, reddish brown. Rays relatively not very distinct without a lens. Pores crowded. Wood moderately heavy.
                                                        RED GUM
  (cc) Heartwood brownish gray. Rays not distinct without a lens. Wood moderately heavy.
    (a3) Pores very small, only occasionally in radial rows of from 3 to 6.
                                                        BLACK GUM
    (b3) Pores slightly larger, often in radial rows of from 3 to 6. (These distinctions between black gum and cotton gum can be applied only by comparison with a piece of wood known to be one species or the other)..............COTTON GUM (TUPELO)
  (dd) Heartwood yellowish brown with greenish tinge. Rays distinct without a lens. Wood moderately light..........YELLOW POPLAR
  (ee) Heartwood creamy brown. Rays distinct without a lens; not in tiers. Wood light...................... ....BASSWOOD
  (ff) Heartwood creamy white. Pores v minute. Rays very fine, barely distinct with a lens; arranged in tiers, producing very fine bands running across the tangential surface. Wood light.
                                                        YELLOW BUCKEYE
                                                        OHIO BUCKEYE

CONIFERS

II. Wood without pores. The cells (tracheids) very small, barely visible with a lens; practically uniform in size, except in the summerwood, where they are narrower radially; and arranged in definite radial rows. Rays very fine.
  A. Resin ducts present but often not distinct without a lens. (Exudations of resin over the end surface are a positive indication of the presence of resin ducts.)
    1. Resin ducts numerous; scattered singly; conspicuous under a lens and usually visible without a lens as minute openings, or more often as darker or lighter colored specks, or as brownish lines on longitudinal surfaces........................The PINES
  AA. Summerwood inconspicuous and not perceptibly harder than the springwood.
                                                        The SOFT PINES:
    (a) Wood soft and moderately light; straight-grained. Annual rings of moderate width. Heartwood light reddish brown.......WESTERN WHITE
                                                        PINE LIMBER PINE
    (b) Wood hard and moderately heavy to very heavy; often cross-grained. Annual rings narrow.
      (aa) Heartwood reddish brown. Tangential surface has numerous slight depressions, which give it a dimpled appearance, especially noticeable on split surfaces......................BRISTLE–CONE PINE
      (bb) Heartwood creamy brown. Tangential surface not dimpled. Wood sometimes very heavy.........................PIÑON (PINE)
  BB. The summerwood conspicuously darker and harder than the springwood. (This feature is not so noticeable in the sapwood of old trees as in the heartwood, where the annual rings are wider.)........................*The HARD PINES:*
    (a) Wood moderately light...................... *WESTERN SPECIES:*
      (aa) The sapwood usually less than 2 in. wide (mostly about 1 in.). Tangential surface has numerous slight depressions, which give it a dimpled appearance, especially noticeable on split surfaces. Resin ducts small, not visible without a lens.............LODGEPOLE PINE
      (bb) The sapwood usually over $2\frac{1}{2}$ in. wide (mostly over 3 in.). Tangential surface rarely dimpled. Resin ducts comparatively large, usually visible without a lens..............WESTERN YELLOW PINE
    (b) Wood moderately heavy to very heavy. Heartwood orange-brown to reddish brown..............*EASTERN SPECIES:* NORWAY PINE
        PITCH PINE   LOBLOLLY PINE   LONGLEAF PINE
        SHORTLEAF PINE   POND PINE   SLASH PINE
    (For distinguishing longleaf from loblolly and shortleaf pine see Fig. 27.)

2. Resin ducts not numerous; scattered singly or in tangential groups of from 2 to 20; not visible without a lens, or appearing as whitish specks in the summerwood.

AA. The heartwood of the same color as the sapwood, or slightly darker, usually not clearly defined. Wood light...................ENGELMANN SPRUCE

BB. The heartwood decidedly darker than the sapwood. Wood moderately heavy.

(a) Heartwood orange-reddish to red. Sapwood over one inch wide.
DOUGLAS FIR

(b) Heartwood russet brown. Sapwood usually less than 1 in. wide.

(aa) Annual rings narrow......................WESTERN LARCH

(bb) Annual rings moderately wide.....................TAMARACK

B. Resin ducts normally absent.

1. The heartwood of about the same color as the sapwood, distinction not clear.

AA. Wood has a spicy odor and taste; moderately light. Summerwood inconspicuous. Color pale brown..................PORT ORFORD CEDAR

BB. Wood does not have a spicy odor or taste, although other characteristic odor may be present.

(a) Wood whitish, at least in the springwood; the summerwood may be dark reddish brown, especially in pieces of rapid growth, and in that case forms a decided contrast between the springwood and summerwood.

(aa) Freshly cut surface of dry wood has a rank odor. Little contrast between springwood and summerwood. Rings of moderate width. Wood light......................................ALPINE FIR

(bb) Freshly cut surface of dry wood does not have a rank odor. Decided contrast between springwood and summerwood. Rings usually fairly wide. Wood moderately light.

(a3) Outer bark containing whitish layers...........WHITE FIR

(b3) Outer bark containing thin, very dark reddish-brown layers.
GRAND FIR

(b) Wood has a reddish hue; even the springwood has a pale reddish color, thus making the contrast between the springwood and summerwood less pronounced; odor somewhat sour in fresh wood. Wood moderately light to moderately heavy.

(aa) Wood coarse and splintery, often cup-shaken. Abnormal resin passages not present......................EASTERN HEMLOCK

(bb) Wood not very coarse or splintery, usually not subject to cup-shake. Abnormal resin passages occasionally present in tangential rows in the outer portion of the summerwood......WESTERN HEMLOCK

2. The heartwood distinctly darker than the sapwood.

AA. Heartwood deep reddish brown. Annual rings regular in width and outline. Sapwood over 1 in. wide. Wood odorless and tasteless; moderately light.
REDWOOD

BB. Heartwood light brown to dingy brown with or without reddish tinge. Odor distinct when fresh surfaces are exposed.

(a) Odor somewhat rancid; heartwood tasteless. Annual rings mostly irregular in width and outline. Sapwood usually over 1 in. wide. Color highly variable from pale brown with or without reddish tinge to blackish brown. Weight variable from moderately light to heavy. Longitudinal surfaces feel and appear waxy.............................BALD CYPRESS

(b) Odor aromatic (like cedar shingles); heartwood slightly bitter in taste. Annual rings narrow but regular in width. Sapwood rarely over 1 in. wide. Wood very light in weight. Longitudinal surfaces not appearing waxy.

(aa) Heartwood brown with reddish tinge...WESTERN RED CEDAR

(bb) Heartwood brown, rarely with reddish tinge.
ARBORVITAE (NORTHERN WHITE CEDAR)

NOTE: The above information and "keys" (Tables II and III) are taken from the *Wood Handbook* and other publications of the Forest Products Laboratory of the U.S. Department of Agriculture. Every rigger should have a copy of this book in his possession, obtainable from the Superintendent of Documents, Government Printing Office, Washington, D.C.

CHAPTER VII

PLANKS FOR SCAFFOLDS

Construction jobs of all types require the use of scaffolds in order that workmen can reach and work in locations which are otherwise inaccessible. The larger the job the bigger the scaffold and the more men who entrust their lives to it. Exclusive of the swinging type of scaffold, such as is used by painters, all scaffolds have a decking constructed of planks, any one of which in falling may be the direct or indirect cause of serious injury or loss of life.

It is therefore not only desirable but essential that only dependable lumber be used for scaffold planks. Eastern spruce is frequently used for this purpose, as its strength is high relative to its weight, while southern pine and Douglas fir are stronger but much heavier.

In certain sections of the country scaffold planks come 2 in. $\times$ 9 in. $\times$ 13 ft 0 in., while in other locations 2 in. $\times$ 10 in. $\times$ 16 ft 0 in. or 2 in. $\times$ 12 in. $\times$ 16 ft 0 in. is standard. For the erection floor on steel buildings under construction 3-in. planks are used because of the heavy loads placed on them.

It is possible to purchase so-called "scaffold planks" from the lumber vendor, but such planks may be of variable or questionable quality. Therefore the purchaser who uses large quantities of scaffold planks should buy them according to specification. It is assumed that the loads can be applied to either the wide or narrow faces of the plank when used as a beam. The plank should have a strength ratio of at least 80 per cent; in other words, the strength of the plank with all its inherent defects should be not less than 80 per cent of the strength of a theoretically flawless plank. This means that the figures published for the strength properties of small specimens of various species of clear, dry wood should be reduced by 20 per cent before dividing by the factor of safety to obtain the safe working load.

SPECIES AND GRADES

The planks should be of the indicated grade of one of the following species as specified in the purchase order:

152

Species	Grade Required
Eastern spruce (*Picea marianna, P. rubra, P. glauca*)	1,200 lb for structural spruce
Sitka spruce (*Picea sitchensis*)	Structural scaffold plank
Longleaf yellow pine (*Pinus palustris*)	Prime structural
Shortleaf yellow pine (*Pinus echanita*)	Dense structural
Douglas fir (*Pseudotsuga taxifolia*)	Select structural

The planks should conform to the standard grading rules of the lumber manufacturers' associations for the grades specified, within the limitations stated hereinafter.

Species	Grading Association
Eastern spruce	Northeastern Lumber Manufacturers' Association
Sitka spruce	West Coast Lumbermen's Association
Longleaf yellow pine	Southern Pine Association
Shortleaf yellow pine	Southern Pine Association
Douglas fir	West Coast Lumbermen's Association

DIMENSIONS

Planks should be unsurfaced, of the following nominal dimensions as called for in the purchase order, and the actual dry dimensions should be not less than indicated below:

Nominal Dimensions	Minimum Actual Dimensions
2 in. × 9 in. × 13 ft 0 in.	$1\frac{7}{8} \times 8\frac{3}{4}$ in.
2 in. × 10 in. × 16 ft 0 in.	$1\frac{7}{8} \times 9\frac{3}{4}$ in.
2 in. × 12 in. × 16 ft 0 in.	$1\frac{7}{8} \times 11\frac{3}{4}$ in.

Actual lengths should be not more than $\frac{1}{4}$ in. shorter nor 1 in. longer than the nominal length.

QUALITY OF WOOD

All planks should be properly seasoned and free from bow, crook, cup, or twist warping.

Boxed heart material should not be accepted.

Pieces of wood exceptionally light in weight for its species and all brash wood should be rejected, as determined by a penknife test.

Flat-grain lumber is preferred for scaffold planks.

DEFECTS

Cross Grain. The slope of the grain, either diagonal or spiral, within the middle half of the length of the plank shall not deviate from a line parallel to the edges of the plank more than 1 unit in a

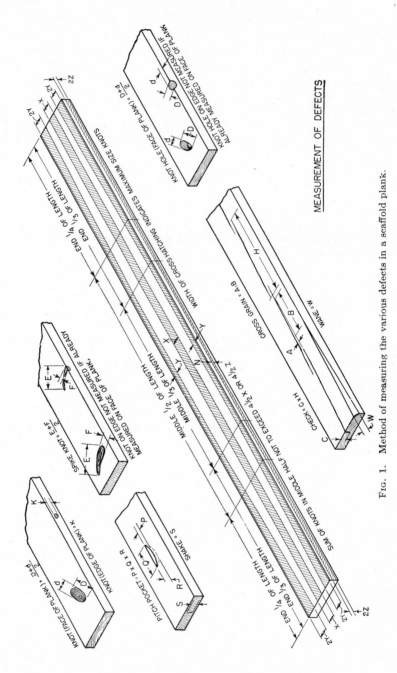

Fig. 1. Method of measuring the various defects in a scaffold plank.

length of 17 units (1:17), as indicated by resin ducts, by splinters pulled from the faces of the plank, or by an ink test. Elsewhere the slope of the grain should not exceed 1:12. Local cross grain caused by knots, etc., shall not exceed 1:12 if within $\frac{1}{2}$ in. of any edge and if within the middle half of the length of the plank.

Knots. Knots appearing within the middle third of the length on the narrow faces should not exceed $\frac{3}{8}$ in. in size.

Knots appearing within the middle third of the length near the edges on the wide faces should not exceed the following limits:

Width of Plank, in.	Max. Size of Knot, in.
9	1
10	1
12	$1\frac{1}{4}$

Knots appearing at the center line of the wide faces shall not exceed the following limits:

Width of Plank, in.	Max. Size of Knot, in.
9	$1\frac{3}{4}$
10	2
12	$2\frac{1}{4}$

Knots appearing on the wide faces between the center line and the edges should be prorated in size according to their position relative to the knots described above (Fig. 1).

The sum of the sizes of all knots appearing within the middle half of the length of 2-in. planks, on any face, should not exceed $4\frac{1}{2}$ times the size of the largest knot allowed in that area, in accordance with the following limits:

Width of plank, in.	Max. total for narrow faces, in.	Max. total for wide faces, in.
9	$1\frac{3}{4}$	8
10	$1\frac{3}{4}$	9
12	$1\frac{3}{4}$	10

Checks. Checks should not exceed 1-in. depth. Length and width are immaterial.

Splits. Splits extending more than 15 in. from either end of a plank should not be permitted.

Shakes. Shakes, as observed on the ends of a plank within the middle half of the width of the wide face, should not exceed $\frac{5}{8}$ in.

measured between lines drawn parallel to the wide faces and just enclosing the shake. Shakes should not appear on the faces of a plank. If both checks and shakes are present, the sum of their sizes at any point should not exceed $\frac{5}{8}$ in.

Wane. The diagonal width of wane on any edge should not exceed $1\frac{1}{8}$ in.

Compression Wood. Compression wood will not be permitted.

Compression Failures. Planks containing or suspected of containing compression failures should not be accepted. Neither should planks with evidence of other injury.

Pitch Pockets. Pitch pockets and bark pockets will be permitted, provided they are not more than 2 in. long, $\frac{1}{8}$ in. wide, or $\frac{1}{2}$ in. deep, and provided they do not occur less than 4 ft apart.

Pitch Streaks. Planks containing pitch streaks of exceptionally large area should not be accepted.

Holes. Knot holes and holes from other causes will be limited as are knots, except that holes should not exceed 1 in. in the minor dimension.

Any planks that meet the requirements of the specifications but about which the inspector may have any doubt can at his discretion be subjected to a load test as described in Chap. VIII. When finally accepted, the plank should be immediately branded on the ends or otherwise permanently marked as suitable for scaffolding. The use of other than "accepted scaffold planks" for scaffold purposes should be forbidden. Inspected planks should be handled carefully and not dropped in bulk from the delivery truck. They should be handled with care during the construction or demolition of the scaffolding and at all other times, as compression failures may be produced when an approved plank is dropped or when a load of unknown magnitude is applied to it.

It is not expected that planks to meet these requirements can be picked from the common grades of lumber; they can be procured only by selecting from the higher grades of lumber. This will undoubtedly increase the cost of the planks, but the additional expense is considered well warranted by the greater uniformity and the greater safety afforded the workmen.

Figure 2 gives the safe concentrated and uniformly distributed loads which may be applied to scaffold planks.

Thought is also being given at this time to the use of metal scaffold "planks." These may be of pressed steel, aluminum, or magnesium having a checkered or other nonslip surface. Such planks may be expected to have many advantages over the conventional wood planks, such as (1) more uniform strength, (2) more readily inspected, (3) lighter weight, (4) fireproof, (5) longer life (as they cannot be cut up and used for blocks or other purposes), (6) will not decay, (7) will not twist warp.   An objection will be the much higher cost, but it is believed that the longer life will warrant it.

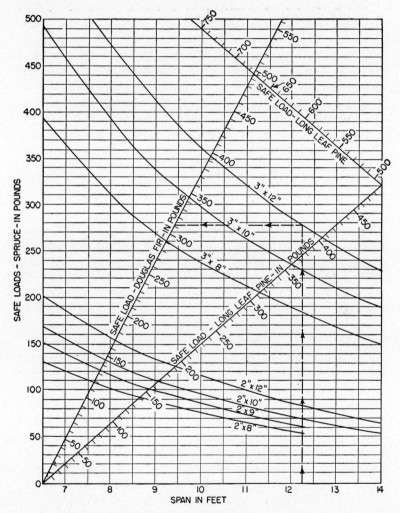

Fig. 2.   Safe concentrated loads on scaffold planks.   For distributed load, double these figures.   (*National Safety Council.*)

CHAPTER VIII

TESTING SCAFFOLD PLANKS

There is an urgent need for a means of subjecting scaffold planks to a load test, which it may be assured will not weaken a good plank yet which may be expected to indicate a defective plank. After several years of research and experiment a method has been developed that, though perhaps not infallible, has proved itself reliable during many years of actual practice by lumber inspectors passing on large quantities of newly purchased scaffold planks.

In some states the authorities require that all scaffolds be tested each time they are erected for use. This may be interpreted to mean that scaffold planks also shall be tested. It is usually specified that the scaffold (or plank) shall be placed a foot above the ground and that three times the normal working load be applied. Such wording of the law or code is rather ambiguous, as it does not indicate whether the load is all to be applied at the middle of the span or is to be uniformly distributed along the length of the plank. A plank or, in fact, any beam whether of wood or of steel will carry a distributed load twice as great as a concentrated load.

Then again, there is a marked difference in the strength of a plank under a steady load and under a suddenly applied load. Thus, the test called for by law is very indefinite. In fact, many engineers believe that a load test on a scaffold plank is not only useless but actually hazardous, for unless properly done there is a far greater probability of injuring a good plank than of detecting a defective one that could not have been detected by a visual inspection alone. For this reason they discourage the testing of scaffold planks except insofar as it is necessary to comply with the law or if there is a reasonable doubt in the mind of the inspector.

The safest method of ensuring that only good scaffold planks are purchased is to have a competent lumber inspector or an engineer who is thoroughly familiar with the behavior of wood under stress make a careful visual inspection of all faces and the ends of each plank, looking for oversize knots, splits, compression wood, compression failures, brash wood, shakes, decay, etc., as described in Chap. VII. However, occasionally after having inspected a

158

plank and having found no defect to which he can point his finger, the inspector may feel in his subconscious mind that there is a question as to its strength. To satisfy himself, he may then desire to apply a load test to the questionable plank. It is only in instances of this kind that a load test should be permitted, and even then it should be conducted only in a scientific manner according to definite rules if it is to be of any value in enhancing the workmen's safety.

Scaffold planks come in different sizes and materials in different parts of the country, 2 in. × 9 in. × 13 ft 0 in., 2 in. × 10 in. × 16 ft 0 in. and 2 in. × 12 in. × 16 ft 0 in. being most common. Although various species are used for this purpose, eastern spruce, longleaf yellow pine, and Douglas fir are most common.

For lack of more specific information, we shall assume that on the job one man weighing 180 lb stands at the center of the plank. Three times this figure would give a proposed test load of about 540 lb. Let us say that a rough (unsurfaced) 2 in. × 9 in. × 13 ft 0 in. eastern spruce scaffold plank is to be tested. In practice, the span is limited to 10 ft, but for test purposes the fulcrums are placed 6 in. from each end, thus giving a span of 12 ft 0 in. Eastern spruce as used in the scaffold plank having a strength ratio of 80 per cent may be expected to have properties approximately as follows:

lb per sq in.

Modulus of rupture (ultimate strength) under static loading...	8,100
Proportional (elastic) limit under static loading..............	5,200
Proportional (elastic) limit under impact loading............	9,150
Modulus of elasticity...................................	1,440,000

It should be noted that the proportional limit under impact is 75 per cent higher than the proportional limit under a slowly applied load; in fact, it is 13 per cent higher than the ultimate strength under static loading. The test stress should not exceed 80 per cent of the proportional limit for static loading, or 4,160 lb per sq in. To develop a bending stress of 4,160 lb per sq in., a concentrated load of about 585 lb has to be applied to the center of the plank, and this will cause the plank to deflect or bend about $5\frac{3}{4}$ in. at the center.

The 2 in. × 9 in. × 13 ft 0 in. spruce plank is placed on two fulcrums $5\frac{3}{4}$ in. high, 6 in. from each end of the plank, on a smooth level floor. A concentrated load of 585 lb will deflect the plank until it just touches the floor and develops a unit stress of 4,160 lb per sq in., as indicated above, and no amount of additional loading can cause any greater deflection or stress. Now if this plank was suddenly deflected 5¾ in., such as by two or three men standing on it and springing it up and down, the stress would likewise be limited to 4,160 lb per sq in. (Fig. 1). But under impact the test

Fig. 1.   Making the "jump test" on a scaffold plank at the lumber dealer's yard.

stress would be only 4,160/9,150 = 45.5 per cent of the proportional limit under impact loading.

In other words, the factor of safety relative to the proportional limit is much higher under the impact test. Therefore, there should be little or no danger of damaging a good plank by subjecting it to an impact test.

TABLE I.   HEIGHTS OF FULCRUMS AND MAXIMUM TEST LOADS FOR SCAFFOLD PLANKS

Species	Size of plank					
	2 in. × 9 in. × 13 ft 0 in.		2 in. × 10 in. × 16 ft 0 in.		2 in. × 12 in. × 16 ft 0 in.	
	in.	lb	in.	lb	in.	lb
Spruce............	5¾	585	9	510	9	632
Longleaf pine*....	6	845	9¼	750	9¼	902
Douglas fir.......	5¾	665	9	598	9	720

* If necessary, longleaf pine planks may be tested on the same fulcrums as spruce and Douglas fir.

On the other hand, wood that is brash or contains compression failures is noticeably weak under impact. Thus, the test recommended here is least likely to damage a good plank, yet most likely to break a plank containing the two most treacherous defects.

It is therefore considered more desirable to apply an impact load to the plank, assuming, of course, that the deflection is limited. Under no circumstance should an impact load be applied to a plank except when the maximum allowable deflection has been mathematically determined in advance. The recommended test should be accomplished by having two or three men stand

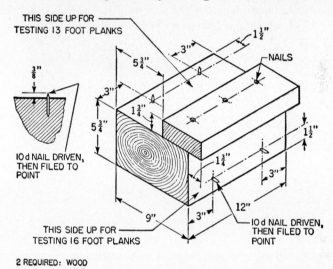

Fig. 2. Details of fulcrum blocks for testing scaffold planks.

close together at the middle of the plank and spring it up and down until it touches or nearly touches the floor several times. The floor, of course, should be a plane surface. If there is no cracking or splitting sound or other evidence of failure under the test, the plank should be turned over and the load applied to the other face.

If there should be a *faint* snapping sound (caused by the wood fibers pulling away from a knot), turn the plank over and apply the test on the other side. Then turn it back, and again test the side that gave the sound. If no further snapping sound is heard, the plank can be accepted, provided, of course, that it has passed the visual inspection (see Chap. VII).

This method has been used for testing such planks as had to be tested by a large user of scaffold planks for over thirty years with very satisfactory results.

Table I gives the height of the fulcrums and the approximate static load to deflect to the floor the three most common sizes of planks of spruce, longleaf yellow pine, and Douglas fir. The actual weight of the men making the test need be only about two-thirds of the static test load indicated in the table. Figure 2 shows a typical fulcrum that can be used for 13- and 16-ft planks of spruce and Douglas fir and, if necessary, for longleaf pine also. The nail points projecting from the blocks will keep the plank from "walking" off the fulcrums as it is sprung up and down.

CHAPTER IX

SWINGING AND SUSPENDED SCAFFOLDS

Swinging Scaffolds. In its most familiar form the swinging scaffold consists of a frame similar in appearance to a ladder with a decking of wood slats and supported near each end by a steel stirrup to which is attached the lower block of a set of manila-rope falls.

The frame is usually constructed of Sitka spruce rails of the depth at the center, as indicated in Table I, which dimension may be reduced by 1 in. at the ends.

The rungs should be of oak, ash, or hickory, at least $1\frac{1}{8}$ in. diameter and spaced not more than 18 in. apart, and the flooring $\frac{1}{2} \times 3$ in. The over-all width of the scaffold is between 20 and 30 in. A 1×4 in. toeboard is provided on the outboard side and, if hinged, is lowered flat against the flooring when the scaffold is in transit. Also on the outboard side is a guardrail not less than 2×3 in. located between 34 and 48 in. (preferably 42 in.) above the flooring, inserted into sockets or loops in the stirrups provided for this purpose. Additional stanchions are provided to keep the span of the guardrail to 10 ft or less. A screen of $\frac{1}{2}$-in. mesh "rabbit wire" is recommended between the guardrail and the toeboard.

Steel stirrups, which support the platform, should be placed between 6 and 18 in. from the ends of the scaffold and secured to it by U bolts of adequate size.

A set of $\frac{3}{4}$-in. No. 1 grade manila-rope falls consisting of a double and a single-pulley block should be provided at each end of the scaffold. A safe means of supporting the upper blocks is absolutely necessary. This may be in the form of roof or cornice hooks, $\frac{1}{2}$-in. wire-rope slings, or other approved device. A tieback rope should be used to secure the cornice hook to a fixed anchorage on the roof.

It is good practice to wire the hook on the lower pulley block to the "eye" on the stirrup to prevent accidental detachment of the hook. A mousing would ordinarily be used in a case like this, but here it would interfere with the hitch of the rope. The hitch is

TABLE I. LADDER-TYPE SCAFFOLDS *

		15 ft 0 in.	16 ft 0 in.	18 ft 0 in.	20 ft 0 in.	24 ft 0 in.
Length of platform	Max.					
Width of platform	Max.	30 in.	30 in.	30 in.	30 in.	30 in.
Cross section of stringers:						
At middle	Min.	$1\frac{7}{8} \times 3\frac{3}{4}$ in.	$1\frac{7}{8} \times 3\frac{3}{4}$ in.	$1\frac{7}{8} \times 4$ in.	$1\frac{7}{8} \times 4$ in.	$1\frac{7}{8} \times 4\frac{1}{2}$ in.
At ends	Min.	$1\frac{7}{8} \times 2\frac{3}{4}$ in.	$1\frac{7}{8} \times 2\frac{3}{4}$ in.	$1\frac{7}{8} \times 3$ in.	$1\frac{7}{8} \times 3$ in.	$1\frac{7}{8} \times 3$ in.
Rungs:						
Number	Min.	10	11	12	13	16
Diameter	Min.	$1\frac{1}{8}$ in.	$1\frac{1}{8}$ in.	$1\frac{1}{8}$ in.	$1\frac{1}{8}$ in.	$1\frac{1}{8}$ in.
Tenons	.....	$\frac{7}{8}$ in.	$\frac{7}{8}$ in.	$\frac{7}{8}$ in.	$\frac{7}{8}$ in.	$\frac{7}{8}$ in.
Tie rods:						
Number	Min.	4	4	4	4	5
Diameter	Min.	$\frac{1}{4}$ in.	$\frac{1}{4}$ in.	$\frac{1}{4}$ in.	$\frac{1}{4}$ in.	$\frac{1}{4}$ in.
Flooring	Min.	$\frac{1}{2} \times 3$ in.	$\frac{1}{2} \times 3$ in.	$\frac{1}{2} \times 3$ in.	$\frac{1}{2} \times 3$ in.	$\frac{1}{2} \times 3$ in.

* Minimum dimensions and maximum spans for ladder type swinging scaffolds (New York State Department of Labor).

made by holding a strain on the rope with one hand and pushing a bight of the slack part of the rope through the inverted V of the stirrup, giving the bight a 180-deg twist and placing it over the bill of the hook. The strain on the "live" part of the rope forces the "dead" part into the V, into which it jams (see Chap. II, Fig. 81). Although a very simple hitch, it is dependable.

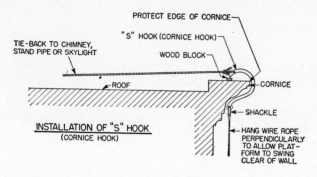

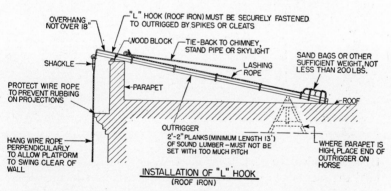

FIG. 1. Roof anchorages for swinging scaffolds. (*Patent Scaffolding Co.*)

To ensure against the scaffold falling in the event the workman should accidentally lose his grip on the rope while raising or lowering the scaffold, a special safety latch can be attached to the cheek of the lower block (Fig. 2). The hauling part, or hand rope, is passed through the hole in the hinged plate, and the hitch made in the usual manner as described above. But should the workman accidentally let go of the rope, the safety latch will raise and grab the rope, thus preventing the scaffold from falling.

Swinging scaffolds are required by law in some states to be not less than 20 in. or more than 30 in. wide. For special jobs, a "box" scaffold shown in Fig. 3 is very satisfactory. The box scaffold provides a two-member guardrail and toeboard on all sides.

When men are working on a swinging scaffold, it must always be secured to the building or structure to prevent it from moving away and allowing the workman to fall between. On building walls it is usually difficult to find something to lash the scaffold to, but it is often found practicable to provide a standard attachment

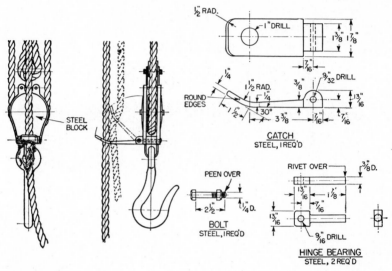

Fig. 2. Safety catch for lower block of rope falls. (*Consolidated Edison Co. of New York, Inc.*)

for a window-cleaner's belt on a short length of manila rope. To hold the scaffold to the wall, the device is attached to the special bolt at the side of one of the windows and the rope secured to the scaffold.

Of course, two hanging life lines are required for all swinging scaffolds, one for each workman. Only new rope, $\frac{3}{4}$ or 1 in. diameter, should be used, and it should be properly secured at the roof or upper part of the structure. At sharp bends over copings or window sills the rope should be padded against abrasion. Each man should wear a 4-in. life belt of three-ply cotton webbing with a $\frac{5}{8}$- or $\frac{3}{4}$-in. rope tail line about 6 ft long (see Chap. XII). The tail line should be attached as short as practicable to the hanging life line by a "rolling hitch" (Chap. II, Fig. 49). This hitch can readily be slid up or down the hanging rope, yet if the man falls,

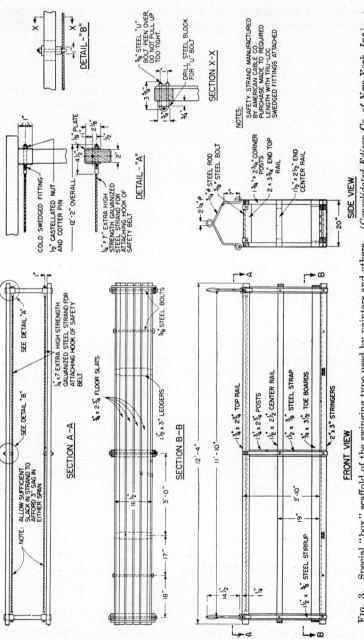

Fig. 3. Special "box" scaffold of the swinging type used by painters and others. (*Consolidated Edison Co. of New York, Inc.*)

the hitch will jam and hold him.   The rolling hitch is similar to the clove hitch except that in tying the first part of the hitch two wraps (instead of one) are made around the hanging line.   Nylon rope may be found suited for life lines, for in addition to being strong it stretches and will stop a falling man more gently.

It should not be necessary to say that the life line must reach to the ground or other place of safety and that the workmen must have their life belts on and attached to the life line at all times, particularly when the scaffold is being raised or lowered.

On the special scaffold (Fig. 3) trolley cables are run along the upper handrail members, and to the cable at the rear the workman attaches the snap hook of his life-belt tail line.   He is then free to walk back and forth in his half of the scaffold.   Should a rope fall fail and one end of the scaffold swing downward, the men probably would remain within the scaffold railings, but in the event they were thrown out, their life belts would keep them from falling to the ground.

In the preceding paragraphs the discussion concerned manila-rope falls for supporting the scaffolds.   It is possible and frequently desirable to substitute wire rope and winches for the manila-rope falls.   Using The Patent Scaffolding Company winch, it is possible to adapt it to the various type scaffolds by resorting to a few bent bars, rods, or simply a little welding.   Figure 4 shows the winch adapted to the box-type scaffold mentioned in the above paragraph. Among the advantages claimed for the winch suspension are the following:

1. Greater safety due to more positive inspections.
2. Ease of handling.   The scaffold can be "inched" up or down as desired.
3. Lower headroom.   The scaffold can be raised closer to the overhead supports.
4. Less danger of failure when acid is used, as in washing building walls.

After the scaffold is erected on a new job, it should be load-tested before men risk their lives on it.   Hoist the scaffold about a foot off the ground, and apply a test weight equal to four men for a period of 5 min.   Tests should also be made every 10 days if the job continues for more than that time.

Where swinging scaffolds are suspended adjacent to each other, planks should never be placed so as to form a bridge between them. Never permit more than two men to work on a scaffold at one time.

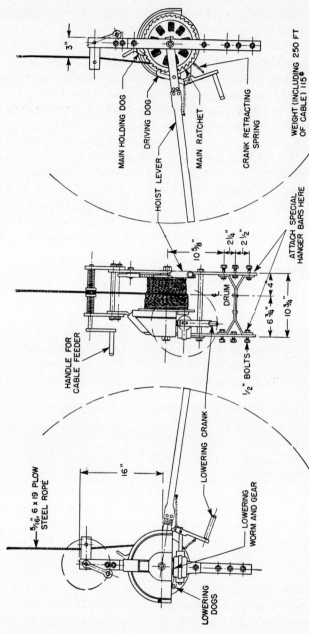

FIG. 4. Patent Scaffolding Co. winch, which can be readily attached to any "box" or swinging scaffold by means of a few special suspension bars.

If the lift of the scaffold exceeds 100 ft, wire rope and winches should be used in place of manila-rope falls. In calculating the strength of swinging scaffolds, a factor of safety of 4 should be used.

Occasionally, it may be desirable to install wood bunters with rollers on them to hold the scaffold away from the building wall and to keep it from swinging or swaying.

When using The Patent Scaffolding Company winches do not under any condition wire the main holding pawl in the disengaged position (Fig. 5). These "dogs" are provided for a reason, and

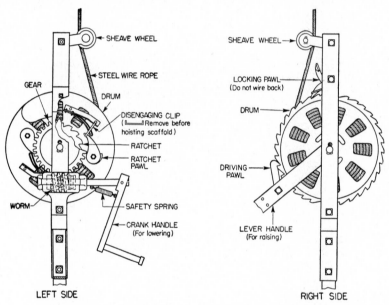

Fig. 5.   The Patent Scaffolding Co. scaffold winch, showing location of two sets of pawls.   (*The Patent Scaffolding Co.*)

to disable them is about as foolish as wiring down a safety valve. In lowering the scaffold it is necessary to hold the dog disengaged, but in case of emergency the workman only has to let go of the dog and allow it to engage the ratchet. The disengaging clips (Fig. 5) that are used to hold the ratchet pawls on the side of the drum in the disengaged position during erection of the scaffold must be removed before the scaffold leaves the ground.

Whenever a winch or scaffold is to be removed to another location, always properly wind the cable on the winch; never coil it up on the ground to save time. A kink in the cable will weaken and

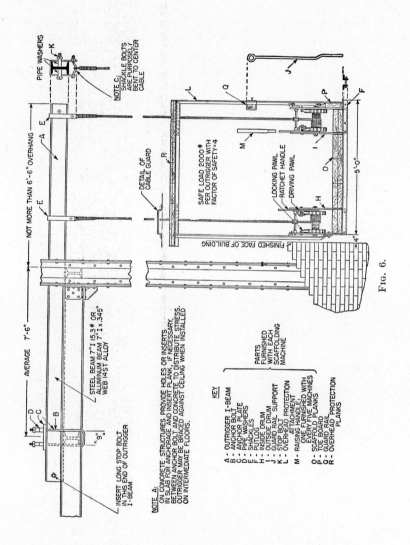

Fig. 6.

PROCEDURE FOR INSTALLING MACHINES

OUTRIGGER I BEAM (*A*). STOP BOLT (*K*)

Install outrigger I beam on roof or floor from which scaffold is to be hung, with stop bolt inserted in hole provided in end of outrigger. Space outriggers at intervals of not over 10 ft from center to center, with overhanging projection of not over 6 ft 6 in. and preferably with outriggers at right angles to face of building.

ANCHOR BOLT (*B*). ANCHOR PLATE (*C*). PIPE WASHERS (*D*)

Clamp outrigger to beam of building or through concrete slab (see Note A) with anchor bolt and anchor plate, using pipe washers on anchor bolt, if necessary, to take up "slack."

SHACKLES (*E*)

Set shackles astride outrigger close to building.

DRUMS (*H*) AND (*I*). PUTLOG (*F*)

Bolt drum frames between the two angle-iron members of putlog, with two putlog bolts resting on bottom of each drum frame. Tighten putlog bolts securely. Alternate holes are provided for putlog bolts in the outside end of putlog, depending on width of plank used for sheathing scaffold. Where 9-in. planks are used, drums are set to allow four planks between them, and with 10-in. planks, drums are set to allow three planks between.

ATTACHING CABLES TO OUTRIGGER I BEAM

Release the cable on drums by temporarily disengaging both locking and driving pawls and pull up cable with light hand line to level of outrigger. Attach eyes in end of cable to shackles. Adjust shackles on outrigger so that shackles will center directly over drums when they are in position on scaffold.

GUARDRAIL SUPPORT (*J*) OR OVERHEAD PROTECTION ATTACHMENT (*L*)

Insert guard rail support through holes in knees on outside drum, or if overhead protection attachment is furnished, attach lower ends of uprights to side of drums. Lugs on outside upright fit in corresponding knees on outside drum, and cotter pin through the lower lug secures the upright to the drum. Slotted knee and one bolt provide means of securing inside upright to inside drum. Enclose cables in cable guards on cross piece of overhead protection attachment. This attachment can be easily unbolted and removed, if desired, at any time during the operation of the scaffold. (*Continued.*)

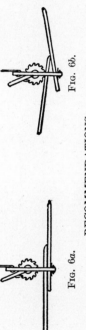

Fig. 6a.

Fig. 6b.

RECOMMENDATIONS

1. Scaffold planks should overlap at least one foot on each side of putlog (Fig. 6a). Overlapping of scaffold planks produces a unique safety element. In the event of catastrophe to cable (Fig. 6b) or overhead support, the planks passing through the drum frame will pinch between the metal putlog and the drum coating, thus sustaining with only slight sagging the weight of the scaffold without support from above.

2. Substantial guardrail or rope should be installed around entire outside of scaffold. (Loop in guardrail support or holes in outside upright of overhead protection attachment provide means of attaching guard rail or rope.)

3. Toeboard should be erected along outside edge of scaffold. (Offset at lower end of guardrail support or on outside upright of overhead protection attachment provides for placing of toeboard.)

4. Wire mesh netting should be provided between guardrail and toeboard.

5. Putlogs must be maintained directly under outrigger I beams, with cables hanging perpendicular; otherwise maximum safety and efficiency will not be obtained.

6. Overhead protection is advisable when work is going on above scaffold.

7. Be sure that structure is sufficiently strong to sustain weight of suspended scaffold or has been adequately shored where necessary.

OPERATION AND INSPECTION OF MACHINES

The machines are raised by a simple lever action and lowered by the reverse of the raising process. To lower, depress ratchet handle with driving pawl held out of engagement, then replace driving pawl and disengage locking pawl, so that weight of scaffold is sustained on ratchet handle, which is allowed to rise, thus unwinding cable from drum and effecting rapid lowering of the scaffold.

There are no delicate, intricate, or concealed parts in the construction of the machines, which have purposely been restricted to the simplest form, as they are usually operated by unskilled labor.

Periodical inspection of scaffolding is recommended while machines are in service. All working parts are visible, so that inspectors and workmen can instantly determine if machines are in safe condition, thus assuring the security of workmen. The cables are subject to constant inspection by workmen while they are being wound on drums as the scaffold is raised, and after having been wound thereon, the cables can neither be damaged nor cause damage.

may even ruin it. The maximum load that can be suspended by one of these winches is 500 lb.

Suspended Scaffolds. Suspended scaffolds are most commonly used by bricklayers on new buildings, but there is no reason why they cannot be used for heavy repair jobs. Hence a few facts will

FIG. 7. Suspended scaffold with overhead protection. Note unsafe overhang of planks at ends of scaffolds.

be mentioned concerning them. This type of scaffold (Fig. 6) consists of a number of outriggers, usually 7-in. steel or aluminum I beams located at the roof level, and from each of which are two $\frac{1}{2}$-in. wire ropes, which wind up on hand-operated winches on the scaffold platform.

The thrust outs or outriggers should not project more than 6 ft 6 in. beyond the point where they bear on the support unless used in pairs. The inboard end should be anchored to the roof steel by large U bolts and anchor plates. These beams should be not less than 15 ft in length and should be spaced not more than 10 ft apart. The suspension cables should be placed not more

Fig. 8.   One-man aluminum safety cage.   (*Patent Scaffolding Co.*)

than 2 and 6 ft, respectively, out from the bearing point of the beam.

The platform consists of planks resting on the putlogs or bearers, which are supported by the winches. The width of the scaffold should not exceed 8 ft. The planking should be $1\frac{1}{4}$ in. thick for spans up to 6 ft and 2 in. thick for spans up to 10 ft. These planks should be laid tight and securely fastened to the putlogs, which they should overlap by not more than 18 in. at each end. A standard guardrail not less than 34 in. or more than

48 in. high and a 9-in. toeboard should be provided along the outer edge of the scaffold. A wire screen between them is recommended.

This type of scaffold equipment provides for a 2-in. plank decking or roofing above the workmen. Special hooks should be provided to hold the scaffold close to the building wall. The allowable loading on an outrigger is approximately 2,000 lb. The platforms on the corner scaffolds (Fig. 7) overhang the bearers by considerably more than the allowable 18 in. and are certainly not considered safe practice.

To raise the scaffold the levers on the winches are operated up and down to rotate them. To lower the scaffold depress the ratchet handle with the driving pawl held out of engagement, then replace the driving pawl and disengage the locking pawl, so that the weight of the scaffold is sustained on the ratchet handle which is allowed to rise, thus unwinding the cable from the drum.

When a relatively minor scaffold job is to be done, a safe and convenient work platform from which to work may be provided by the one-man aluminum safety cage shown in Fig. 8. This lightweight rig is more comfortable for the workman than any boatswain's chair, yet it is nearly as flexible in operation.

CHAPTER X

SCAFFOLDING

Built-up Scaffolds. So-called "built-up" scaffolds may be either of the single-pole type, one side of which is supported by poles or uprights and tne other side by the wall or structure against which it is erected, or of the double-pole type, which is erected independently of the building or structure. Scaffolds of these types require expert workmanship in their construction and hence should not be undertaken by inexperienced men.

Built-up scaffolds are required to be built in conformance with state labor laws, industrial codes, local city ordinances, and national safety codes, all of which are in agreement in most important aspects but which differ in details. Many of the data contained in this chapter are taken from the Industrial Code Rules (No. 23) Relating to the Protection of Persons Employed in the Erection, Repair and Demolition of Buildings or Structures adopted by the Department of Labor of the state of New York. Scaffolds built to meet these requirements should be acceptable in most states and cities.

If any portion of a scaffold has been weakened or damaged by storm, accident, or otherwise, it should not be used until the necessary repairs have been made. Care should be taken to avoid overloading any scaffold. When the allowable uniform load is given as, say, 50 lb per sq ft, it means that a 50-lb load can be applied simultaneously to each square foot of deck area. Naturally a man weighs more than 50 lb, yet he can stand on 1 sq ft of flooring, but it is not expected that men will be crowded on a scaffold. In other words, the load should *average* 50 lb per sq ft. Where there is danger of men at work on a scaffold being struck by material or tools dropped by workmen above, a tight decking or roof of 2-in. planks should be erected above the men.

Every scaffold erected 6 ft or more above the ground must have a guardrail and toeboard along the unprotected edges and ends of the work level. On unprotected scaffolds at high elevations the men frequently wear life belts properly anchored to some substantial part of the structure. Sometimes life nets are provided below

177

to catch a man should he fall from the scaffold. Guardrails must be at least 34 in. high, preferably 42 in., and must be supported every 10 ft or less. If there is danger of material falling from the scaffold, a $\frac{1}{2}$-in. mesh wire screen of No. 18 gauge steel should be provided between the toeboard and guardrail.

Spruce, fir, Douglas fir, and southern yellow pine are most commonly used for scaffold work. Spruce has the highest ratio of

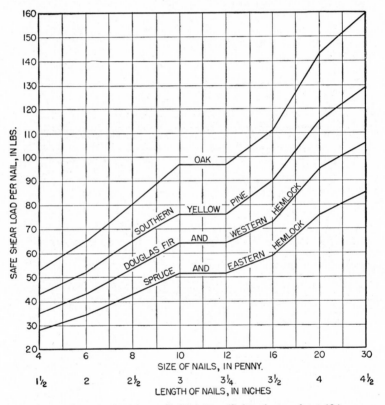

FIG. 1. Safe loads on nailed joints. (Safety factor about 10.)

strength to weight and is therefore often preferred. Lumber for scaffold members and planking should be reasonably free from serious defects such as were described in Chaps. VI and VII. A sufficient number of proper size nails, driven in fully, should be used at each joint or splice in the scaffold. Never use nails in tension, such as when the two members have a tendency to pull away from each other. Rather have the nails subject to shearing stress, such as when the members tend to slip on each other. To develop full strength, at least one-half the length of a nail must be driven into the main member to which the secondary member is being nailed. Figure 1 gives the allowable or safe load per nail for different size nails driven into various kinds of wood. The minimum number of proper size nails for securing a board or plank 4 in. wide is two, for 6 in. wide three nails, 8 in. wide four nails, and 10 or 12 in. wide five nails.

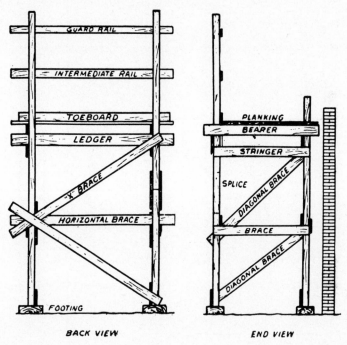

Fig. 2. Double-pole or independent-pole type of scaffold. (*National Safety Council.*)

It is very important that the scaffold uprights or poles be plumb, and they must rest on a proper footing; otherwise as the load is applied to the scaffold, the poles would embed themselves in the ground. Where it is necessary to splice poles, the squared end of the upper member should rest firmly upon the squared upper end of the lower member. To secure these members together, two wood cleats 36 in. long should be nailed on adjacent faces of the poles at the joint. The total cross-sectional area of the two cleats should be at least one-half the cross-sectional area of the pole.

TABLE I. DIMENSIONS OF WOODEN INDEPENDENT POLE SCAFFOLDS

Type of use	Light duty	Regular duty	Heavy duty
Safe uniformly distributed load, max.	25 lb per sq ft	50 lb per sq ft	75 lb per sq ft
Height scaffold and minimum-size poles	20 ft 2 × 4 in. {40 ft 3 × 4 in. {40 ft 2 × 6 in. 60 ft 4 × 4 in. 80 ft 4 × 6 in.	20 ft 3 × 4 in. 20 ft 2 × 6 in. 40 ft 4 × 4 in. 60 ft 4 × 6 in. 80 ft 6 × 6 in.	20 ft 4 × 4 in. 40 ft 4 × 4 in. 60 ft 4 × 6 in. 80 ft 6 × 6 in.
(Lower 20 ft)			
Pole footing, min.	2 × 9 in.	2 × 9 in.	2 × 9 in.
Longitudinal spacing of poles:			
With 1¼ × 9 in. ledgers, max.	6 ft 0 in.		
With 2 × 9 in. ledgers, max.	10 ft 0 in.	10 ft 0 in.	6 ft 0 in.
Transverse spacing of poles, max.	10 ft 0 in.	10 ft 0 in.	10 ft 0 in.
Ledgers (on edge), min.	1¼ × 9 in.	2 × 9 in.	2 × 9 in.
Vertical spacing of ledgers, max.	7 ft 0 in.	6 ft 0 in.	4 ft 6 in.
Bearers (on edge), min.	1¼ × 9 in.	2 × 9 in.	2 × 9 in.
Nonsupporting stringers, min.	1 × 4 in.	1¼ × 4 in. or 1 × 6 in.	1¼ × 9 in.
Bracing, min.	1 × 4 in.	1 × 6 in.	1 × 6 in.
Plank decking:			
Span 6 ft 0 in. or less, min.	1¼ × 9 in.	1¼ × 9 in.	2 × 9 in.
Span more than 6 ft 0 in., min.	2 × 9 in.	2 × 9 in.	2 × 9 in.
Guardrail, min.	2 × 4 in.	2 × 4 in.	2 × 4 in.
Height guardrail, min.	34 in.	34 in.	34 in.
Toeboard, min.	1 × 8 in.	2 × 9 in.	2 × 9 in.

Figure 2 shows a typical wooden independent pole scaffold with the various parts identified. It should be noted that the splices of the poles are staggered. The minimum sizes of members and the maximum spans for light-duty, regular-duty, and heavy-duty scaffolds are given in Table I. The ledgers (the horizontal members extending lengthwise of the scaffold) should never be spliced

Fig. 3. Wood-pole scaffolding contrasted with more modern steel scaffolding. (*Consolidated Edison Co. of New York, Inc.*)

between poles but should always overlap the poles by at least 4 in. Wood blocks are usually nailed to the poles below the overlapped ledgers to aid in supporting them. The bearers or putlogs (the horizontal members extending crosswise of the scaffold) should rest on the ledgers and be nailed to the poles. They should be long enough to project over the ledgers beyond the outer row of poles by at least 12 in. and beyond the inner (toward the building) row of poles by at least 2 in. (Fig. 2).

Diagonal bracing should be provided for the entire outside face of the scaffold. In erecting built-up scaffolds, whether of wood or tubular steel, it is of utmost importance that they be adequately braced against collapse, both lengthwise and crosswise of the scaffold. If the independent pole scaffold is free standing (that is, away from a building or structure and not attached to it), its width must be at least one-third of its height.

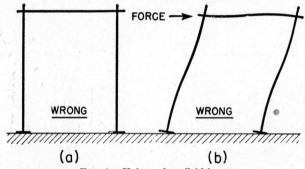

FIG. 4. Unbraced scaffold bent.

The decking planks should be laid with their edges close together so that tools and material cannot fall through. The scaffold planks should overlap each other at the bearers by not less than 36 in., or in other words they must project at least 18 in. beyond the bearers.

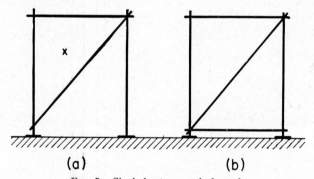

FIG. 5. Single bent, properly braced.

All structures, in order to be ensured against possible collapse due to horizontal forces such as those caused by the wind, moving loads, etc., should have bracing installed in such a manner as to create a number of triangles. The frame, or "bent," shown in

Fig. 4*a* is *not* self-supporting. If a horizontal force or thrust is applied, the frame will assume the shape shown in Fig. 4*b*, which for clarity is greatly exaggerated.

In order to prevent this folding up, a diagonal brace should be installed to form the triangle X shown in Fig. 5*a*. There is still

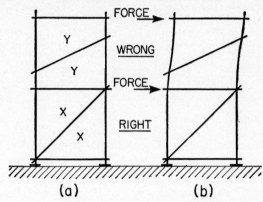

Fig. 6. Proper and improperly placed braces.

the remote possibility of the other pole being bumped into and the lower end displaced, which, of course, would cause collapse, so it is customary to add the horizontal brace shown in Fig. 5*b*, unless both posts are secured to a sill piece.

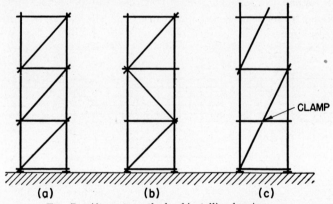

Fig. 7. Alternate methods of installing bracing.

Attention is called to the desirability of forming true triangles as shown in the lower panel in Fig. 6*a*. In the upper panel, the diagonal brace does not intersect the pole and ledger or the bearer in a common point, so when a thrust is applied (Fig. 6*b*), the lower panel will resist distortion while the upper panel, which is inadequately braced, will be deflected as shown. It is not always practicable to install the bracing properly, owing to the need for avoiding obstruction of passageways.

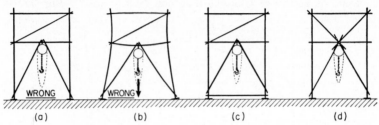

(a) (b) (c) (d)

Fig. 8. Additional bracing may be required to support a concentrated load.

As higher scaffolds are built, additional panels of two triangles each are added. The diagonal braces may all be arranged parallel (Fig. 7a), or they may be placed in a zigzag manner (Fig. 7b). Or if so desired, a long diagonal brace may extend two panels high (Fig. 7c) provided it is secured to the intervening horizontal member.

Where concentrated loads are to be supported, the member or members supporting the load should be properly reinforced or braced. Figure 8a shows an improper arrangement sometimes used. When the load is applied (Fig. 8b), the horizontal member will bend under the weight and the feet of the poles will spread. To overcome this a tie member should be provided (Fig. 8c) to

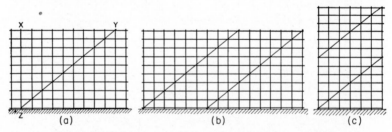

(a) (b) (c)

Fig. 9. Long, continuous bracing members for large scaffolds.

hold the feet of the poles from spreading, or the necessary bracing (Fig. 8d) should be added to prevent the bending of the horizontal member under the load.

On large scaffolds that are many panels wide by many panels high, a continuous diagonal brace may be used. Figure 9a shows

..diagram of a large scaffold along the face of a big building.   The
diagonal in this case, as in every other case, should start at the
ground level and not one panel above the ground as is occasionally
done.   If the scaffold is extra long with respect to its height, two
or more diagonals may be used, placed either parallel (Fig. 9*b*) or
in an inverted V.   If the structure is higher than wide, bracing
may be arranged as shown in Fig. 9*c*.   Note
that the upper brace starts at the same elevation as the top of the lower brace.   On all
such scaffolds, diagonal bracing in a plane at
right angle to the plane shown should be
provided at every second pole and should
continue from the ground to the top of the
scaffold.

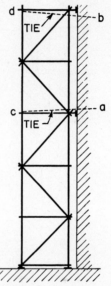

FIG. 10.   Anchoring a
tall, narrow scaffold.

Where the height of the scaffold exceeds
three times its width, the structure should
be secured against overturning bodily.   This
can be accomplished by means of $\frac{1}{2}$-in. steel
cables attached to the outer poles at *c* and *d*
(Fig. 10) and extending at about a 45-deg
angle in a horizontal plane to building columns or other adequate supports *a* and *b*.
Such tie-ins should be provided at every
second or third panel vertically.

Where the scaffold is placed against or
adjacent to an irregular-shape wall (Fig. 11*a*
and *b*), the vertical load from the upper
sections of poles should be transferred to the lower sections by
using the necessary diagonals.   Even under such conditions the
primary system of bracing of the scaffold should be kept continuous.   Figure 11*c* shows a balcony bracketed out from a scaffold.
The brace in this case should be clamped also to the projecting
horizontal member.

Figure 12 shows a gantry built across the entrance to a public
building where additional poles would be objectionable.   In this
case the entire structure must be subdivided into triangles to form
the necessary truss.

Where built-up scaffolding is used inside a room, and where it
extends from wall to wall, the diagonal braces can be omitted provided the scaffold is wedged against the walls (Fig. 13) to prevent
its collapse.   These drawings show the bracing in one plane only.

It should be thoroughly understood that bracing must be provided in the other plane also.

Tubular steel scaffolding, which may be rented or purchased, is to a large degree superseding wood scaffolding, except for the

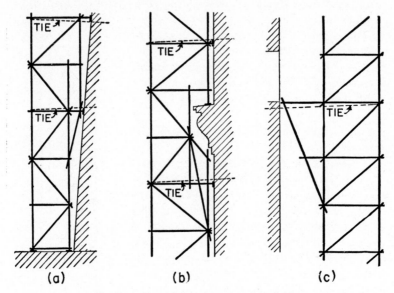

FIG. 11.   Special arrangements of scaffolds.

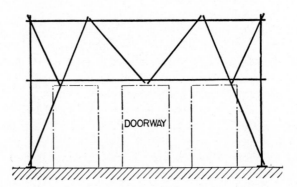

FIG. 12.   For long spans trusses must be constructed.

very smallest jobs, owing to its greater strength and safety and the ease with which it can be erected and dismantled.   Also the members do not deteriorate as do planks and stringers, which warp, split, and decay.   The framework of the steel scaffolding, of course, is noncombustible, and when fireproofed planks are used, the danger of a major fire such as took place in the Sherry-Netherlands tower in New York a number of years ago cannot occur.

On all jobs, but especially on outdoor scaffolding where there is danger of the wind dislodging the planks, they should be secured to the putlogs by means of the special anchors shown in Fig. 14.

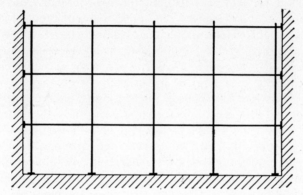

Fig. 13.   No bracing is required where the scaffold abuts the walls inside a room.

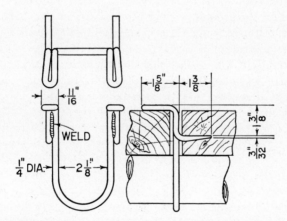

Fig. 14.   Special device for holding scaffold planks against displacement.

When dismantling the scaffold do not drop the planks or the metal parts to the ground.   Careless handling of scaffold members may damage them, and indirectly result in their failure when used on the next job.

The New York Board of Standards and Appeals requires that

legible signs be posted not more than 50 ft apart on the work level to indicate the maximum live load per square foot which may be placed on the scaffolding.   For access to all scaffolding more than 10 ft in height fixed ladders (metal or wood) of standard dimensions should be installed.

Prefabricated Scaffolding.   There are two general types of tubular steel scaffolding, namely the prefabricated type intended for light duty, and the custom-built type intended for special jobs and heavier loads.   The prefabricated type (Fig. 15) consists of light-weight tubular steel frames, which are essentially two short poles, a putlog, and bracing members welded together. It is said that this design was conceived by an engineer who, while confined to a hospital, spent many weeks looking at the "head" and "foot" portions of the beds and wondering why they could not be placed on top of each other to form scaffolding.

The illustration shows the general idea of erecting the prefabricated scaffold.   Two frames are stood up opposite each other on the floor or ground, and the removable cross braces between them are placed on the stud bolts which are welded to, and form an integral part of, the frame.   Of course, the poles or posts of the bottom frames rest on steel bases nailed to planks to distribute the load on the ground.   After the first tier of panels is erected, the second tier of frames are socketed onto their upper ends.

If the illustration of this type of scaffolding (Fig. 15) is examined carefully it will be observed that the diagonal and horizontal (or lack of horizontal) bracing fails to produce true triangles as previously described.   With such slender members as these the strength of the scaffolding is much less than it would be if the bracing was arranged in the ideal manner, for if a horizontal force is imposed on the scaffold, such as by a strong wind, there will be a tendency to develop reverse bends in the posts, as indicated in Fig. 6b.   To resist such wind stresses the scaffolding should be anchored to the building or structure by $\frac{3}{8}$-in. wire rope every 24 feet vertically and every 28 feet horizontally.   In addition, due also to the slenderness of the members, they may be accidentally bent while in transit, and a very slight bend may cause a serious reduction in strength.

The standard frames of various manufacturers are 5 ft wide, and they range in height from 3 ft to 10 ft, 5 ft being most common.   Due to the numerous designs of frames the allowable load uniformly distributed on the putlog (the top member of the frame) varies considerably. Likewise, due to the variation in the length of the posts between the removable cross braces, there is a wide range in the loads which they can support.

The frames are all interchangeable so, unless the rigger knows exactly the load which each model will safely carry, we must limit the live loading to that permitted on the model having the lowest strength.

It is the opinion of the safety engineer that the safety factors ordinarily used for scaffolding are much too low. Even though the dead load of the structure of a building can be figured with a

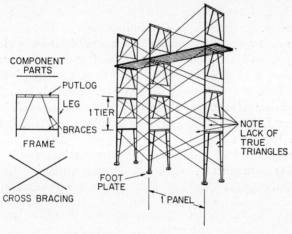

FIG. 15. A typical prefabricated tubular steel scaffold in the process of erection.

high degree of accuracy, and the live load can be closely estimated, building codes usually require a safety factor of between 3 and 4. On the other hand, the dead load on the lower sections of a built-up scaffold varies almost directly with the height to which it is erected. The live load is strictly a guess.

For instance, when a scaffold is about to be erected it may be assumed that one pallet of bricks will be placed on the decking in each panel. But through ignorance or carelessness two pallets are placed in the panels. This will almost double the live load. When the dolly used to move the pallets sets down the load it may produce an impact load of perhaps 100 per cent additional. So, if a safety factor of only $1\frac{1}{2}$ to $2\frac{1}{2}$ had been allowed on the scaffold, as is permitted by some municipal regulations, the scaffold would be in danger of collapse.

The factor which usually limits the loading on a scaffold frame is the strength of the putlog or top horizontal member of the frame, upon which the planks rest.   The total load on any frame is one-half of the live and dead loads on the two adjacent spans of decking.   It should not exceed 2,000 lb on frames having diagonal bracing members, or 1,600 lb on those having no diagonal braces.   These figures are based upon a safety factor of 5.   Two-inch scaffold planks come in 13′ 0 in. lengths, so regardless of the spacing of the frames the same weight of planking will usually be used, the amount of overlap varying as required.   Thus the weight of planking on each frame putlog will be about 250 lbs. The weight of the frames and their removable bracing may be estimated at about 10 lb per foot of scaffold height.

As an example, let us assume that a scaffold is built of frames of the type *having* diagonal braces.   The frames are spaced 8 feet apart (the panel length is 8 feet), and the scaffolding is 120 feet high.   There are two levels decked over with 2-in. planks; one deck is being used, the other is not.   What is the safe live load per square foot on the decking being used by a contractor cleaning the building walls?

From Table II we first investigate the maximum safe load on the putlog of the frames which have diagonal braces, and we find that it will support 1,750 lb.   Following down this column in the

TABLE II.   SAFE LOADS ON PREFABRICATED SCAFFOLD FRAMES

Type of frame				
Load on putlog of frame				
Gross load, including planking......	2,000 lb		1,600 lb	2,000 lb
Live load, exclusive of planking.....	1,750		1,350	1,750
Safe live load per sq ft of planking*				
10-ft panels.......................	40 lb		32 lb	40 lb
9-ft panels........................	44		36	44
8-ft panels........................	50		40	50
7-ft panels........................	57		46	57
6-ft panels........................	67		53	67
5-ft panels........................	80		64	80
Maximum safe load on lowest frame...	6,200 lb		6,200 lb	5,600 lb

* Safety factor = 5.

table we find that with 8-foot panels the safe unit live load on the decking is 50 lb per square foot.

Next we check the strength of the legs of the frames. Naturally, the lowest frame supports the greatest load. From Table II, we find that the gross load on the frame at the decked level is 2,000 lb. This includes the live load and the planking. There is also the unused decked level which supports only the 250 lb of planking. Assuming that the frames weigh about 10 lb per foot of height, the weight of all the frames above the base is 120×10 lb $= 1,200$ lb. So, adding the 2,000 lb, 250 lb, and 1,200 lb, we have a total load on the two legs at the bottom frame at 3,450 lb. From the bottom line in Table II we find that the frames will support 6,200 lb, this being far in excess of the load which is to be applied. A floor load of 50 lb per sq ft is, therefore, safe for the intended service.

Oh, yes, we see dozens of scaffolds which, due to their excessive height, are loaded far beyond this figure. However, just because they have not yet collapsed does not prove that they are safe for the men at work on them, or for the hundreds or perhaps thousands of persons who must walk under or within collapsing range of the scaffolding.

Fig. 16. A view of a prefabricated scaffold of extreme height and badly overloaded. Note that some frames, especially at the ends of the scaffolding, support up to eight decks of planking, some containing masonry removed from the cornice. Also, note the extra row of frames at the upper stories cantilevered out from the body of the scaffolding and thereby adding their weight to the load on the structure below. (Some frames were painted white and some gray, hence the contrast in the photo.)

Figure 16 shows a typical prefabricated scaffold of large size and height, supporting numerous work levels. At both ends of the scaffolding, below the catch-scaffold decking, one will note that a bracket is extended outward to support an additional stack of frames, which of course imposes additional loading on the frames below. This type of scaffolding warrants a careful check-up as to the live load which it can safely support. On all such installations it is recommended that a licensed professional engineer be called upon to check the loading.

TUBULAR SCAFFOLDING

For scaffolding which must have panels and tiers of varying dimensions, or which must fit around obstructions, the Tubelox type is best suited (Fig. 17). To give additional strength the

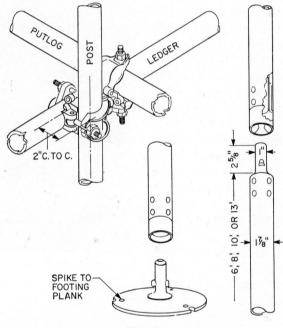

Fig. 17.

posts can be spaced more closely, the length of the putlogs may be shortened, and more adequate bracing may be provided, as in scaffolding standing free of any structure which might support it.

This scaffolding consists usually of members of $1\frac{1}{2}$ in. standard weight galvanized pipe which comes in 6-, 8-, 10-, and 13-foot lengths. The members may be used interchangeably for posts, putlogs, ledgers, bracing, and handrails. The posts are joined end to end by placing the members with the female end down, and locking them to the male upper end of the posts below, by giving a 90-degree twist. For securing the various members together, rigid 90 degree couplers are generally used. Swivel couplers are available where necessary.

In erecting this type of scaffolding the posts are placed on the steel bases which distribute the load on the footing planks. Then the ledgers are clamped to the posts at the desired height. The putlogs or bearers are clamped to the ledgers close to the posts. The braces on the outboard face are clamped to the projecting ends of the putlogs, close to the posts. Cross braces are installed at every fourth or fifth pair of posts. Usually the putlogs extend so as to bear against the wall of the building or other substantial support. Ties of $\frac{3}{8}$-in. wire rope hold the scaffold to the structure as described for prefabricated scaffolds.

As the scaffold is extended upward as construction progresses, the planking can be removed from the lower levels and placed higher up, but the handrail members should remain to give added rigidity to the scaffold.

The post splices should be located a short distance above the ledger couplings, being staggered so as to occur alternately at different levels. Where longitudinal diagonal bracing is made continuous from the ground to the top of the scaffold, perhaps it may not meet the putlogs at all levels; in such cases the braces should be clamped to the post by means of swivel couplers.

To check the strength of tubular steel scaffolding of this type (Fig. 17), first determine the live or movable load which may be applied to the plank decking. This should be not less than 25 lb per sq ft of deck, preferably not less than 50 lb per sq ft. In a scaffold or staging more than one panel in length each putlog carries a load of the decking equal to one-half the panel either side of it, or usually the area of one panel. For instance, if the posts are spaced 5 ft. 0 in. × 5 ft 0 in. the area supported is 25 sq ft. This area, multiplied by 25 lb per sq ft, gives a load of 625 lb uniformly distributed over the length of the span of the putlog.

To this must be added the weight of the plank decking and toeboards, say 240 lb, making a total of 865 lb on the putlog. Referring to Fig. 18, we enter the bottom of the chart at 64-in. (actual span of the putlog), follow up to the curve for $1\frac{1}{2}$-in. pipe, and read the allowable safe load 800 lb. While this indicates a slight overload, the factor of safety will permit us to thus exceed the nominal safe load.

The vertical load on the posts is maximum at the ground level. As each putlog is supported by two posts, the load on each post is one-half that on the putlog, or 433 lb in this case.   If two or more

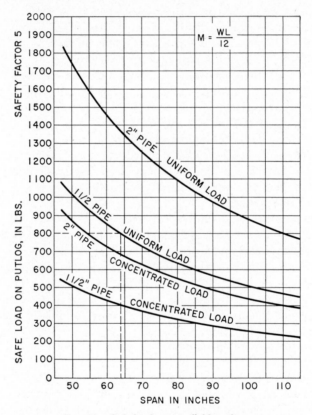

Fig. 18.   Safe loads on scaffold putlogs.

levels are planked over, the load on each post is equal to one-half the loads on all the putlogs above, plus (in the case of very high scaffolding) half the weight of the posts, putlogs, ledgers, railings, etc. above.   This figure should not exceed the safe load on the post obtained from Fig. 19.   Assuming that the height of the posts between ledgers is 6 ft, 6 in., we will enter the bottom of the chart at 78 in., follow upward to the curve for $1\frac{1}{2}$-in. pipe, and

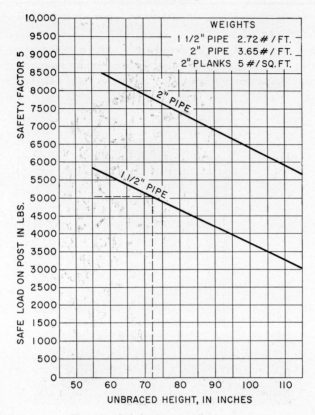

FIG. 19. Safe loads on scaffold posts.

read 5,050 lb safe load at the left margin. Obviously, this is amply safe for supporting several decks.

Portable Scaffolding. For painting interior ceilings, servicing lighting fixtures, and similar jobs a portable all-aluminum tubular scaffolding on rubber tired casters is very handy (Fig. 20). The horizontal and diagonal bracing members, while not in the same plane and therefore not theoretically of proper design, nevertheless have adequate safety for the limited height to which the scaffold is erected. The diagonal braces are so designed as to be used as stairs. This scaffold can be erected or dismantled by two men in a relatively few minutes. The casters should be locked against movement before anyone goes onto the scaffold. If the scaffold is more than three panels high it should be secured against upset by outriggers, lashings or guys.

Fig. 20.   Portable aluminum scaffolding for use indoors.   (*Chesebro-Whitman Co., Inc.*)

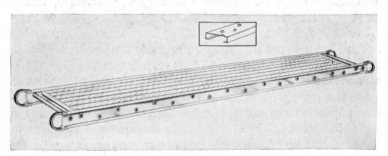

Fig. 21.   Aluminum staging for scaffold.   (*Werner Aluminum Co.*)

Material Hoist Towers. Towers for platform-type material hoists are also constructed of tubular steel (Fig. 22), but of heavier members. These towers can be erected to great height, such as at tall buildings under construction, but they must be properly guyed or anchored to the buildings they serve. There is one point to be kept in mind; diagonal braces should be used to form triangles on all four faces of the tower. This includes the loading and unloading sides where the bracing is commonly omitted to permit handling of material at the landings.

Where cross bracing is impracticable, then by all means provide knee braces at the upper corners of the landing openings to afford at least some rigidity. Figure 9 in Chapter XXI shows quite vividly how the lack of diagonal bracing on the landing side of a hoist tower permitted the collapsing staging to pull the hoist tower over with it.

Outrigger Scaffolds. This type of scaffold is frequently projected from a window of a building to enable workmen to reach the upper portions of the wall or cornice. Timbers 3 × 10 in., used on edge, are placed horizontally, resting on a piece of plank placed on the window sill, and extend not more than 6 ft beyond the face of the wall. To resist the upward thrust at their ends inside the building, they are usually nailed to planks placed vertically and wedged against bearing blocks at the floor and ceiling. Cross bridging between the outriggers at the window sill resists any tendency of the planks to roll over.

The maximum spacing between outriggers is 10 ft. On the outriggers are nailed 2 × 9 planks laid close together, although a 3-in. space at the wall is permitted when necessary. The end planks should not extend more than 18 in. beyond the last outriggers.

Standard guardrails and toeboards, as described above, are required along the outboard edge and ends of the scaffold, but in erecting the guardrails special attention should be paid to making the stanchions of adequate strength. Screening of $\frac{1}{2}$-in. square mesh is desirable between the guardrail and toeboard.

Needle Beam Scaffolds. This type of scaffold is used only for very temporary jobs, particularly for working on steel structures. No material should be stored on this scaffold. Two needle beams, 4 × 6 in. on edge, are placed parallel to each other in a horizontal or nearly horizontal plane (depending upon circumstances) and nor more than 10 ft apart. They are usually suspended by rope near each end, and a center support is always required. The rope should be 1-in. No. 1 grade manila or larger and is attached to the needle beams by a scaffold hitch (Chap. II, Figs. 77 to 79) so as to prevent the beams from rolling over on their sides under load.

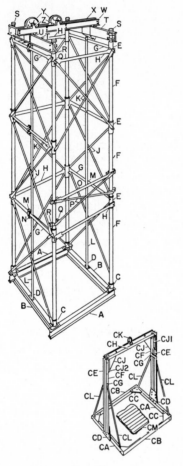

FIG. 22.

The rope should then extend up over a structural member, down under the needle beam, and back over the structural member, after which it is secured. The hitch shown in Chap. II, Fig. 80, is frequently used, although a clove hitch or rolling hitch is preferable. Precautions must be taken to prevent the ropes from slipping off the ends of the needle beams, particularly if the latter are inclined.

The decking should be of 2-in. planks having a length at least 2 ft longer than the span between needle beams, and the length of the decking in a direction parallel to the beams should be not less than 3 or more than 6 ft. If the scaffold is not level, the planks should be nailed to the beams or have cleats nailed on their undersides to engage the beams.

In constructing or using wood scaffolds and planking, special attention should be paid to the condition of the wood, as described in Chaps. VI and VII.

Where local conditions necessitate a long span between scaffold supports, the prefabricated aluminum staging (Fig. 21) may be used to good advantage. In fact, there are prefabricated scaffold assemblies which provide a long platform of similar type on adjustable legs.

SINGLE AMERICAN TUBULAR TOWERS (Fig. 22)

(Two-wheelbarrow Cage)	(Three-wheelbarrow Cage)

BASE

Letter	Description
A—Base Sill I Beam—7 ft. 2 in. long
B—Base Sill I Beam—6 ft. 0 in. long
C—Base Casting
D—Bottom Pipe Guide Support

BASE

Letter	Description
A—Base Sill I Beam—8 ft. 10 in. long
B—Base Sill I Beam—6 ft. 0 in. long
C—Base Casting
D—Bottom Pipe Guide Support

INTERMEDIATE SECTIONS

E—Sleeve—11 in. long
F—Leg—6 ft. 6 in. long
G—Girt—6 ft. 6 in. long
H—Girt—6 ft. 6 in. long
J—Brace—9 ft. $2\frac{15}{16}$ in. long
K—Brace—9 ft. $2\frac{15}{16}$ in. long
L—Guide—2 in. Pipe 6 ft. $4\frac{7}{8}$ in. long
M—Guide Insert (Malleable)
N—Guide Clamp (Forged Steel)

INTERMEDIATE SECTIONS

E—Sleeve—11 in. long
F—Leg—6 ft. 6 in. long
G—Girt—6 ft. 6 in. long
H—Girt—8 ft. 2 in. long
J—Brace—9 ft. $2\frac{15}{16}$ in. long
K—Brace—10 ft. $5\frac{11}{16}$ in. long
L—Guide—2 in. Pipe 6 ft. $4\frac{7}{8}$ in. long
M—Guide Insert (Malleable)
N—Guide Clamp (Forged Steel)

LANDING PANELS

O—Landing Panel Girt—2 in. × 6 ft. 6 in.
P—Landing Panel Brace—1 in. × 7 ft. $2\frac{1}{8}$ in.
Q—Landing Panel Clamp—Front Half
R—Landing Panel Clamp—Back Half

LANDING PANELS

O—Landing Panel Girt—2 in. × 8 ft. 2 in.
P—Landing Panel Brace—1 in. × 8 ft. $8\frac{1}{2}$ in.
Q—Landing Panel Clamp—Front Half
R—Landing Panel Clamp—Back Half

TOP AND CATHEAD

S—Top Leg
T—Top Channel—6 ft. 6 in. long
U—Cathead Clamp (Malleable Casting)
V—Cathead U-Bolt
W—Cathead Channel—11 ft. 4 in. long —Right Hand
X—Cathead Channel—11 ft. 4 in. long —Left Hand
Y—Sheave Wheel
Z—Cathead Bearing

Q—Landing Panel Clamp—Front Half
R—Landing Panel Clamp—Back Half

TOP AND CATHEAD

S—Top Leg
T—Top Channel—6 ft. 6 in. long
U—Cathead Clamp (Malleable Casting)
V—Cathead U-Bolt
W—Cathead Channel—11 ft. 4 in. long —Right Hand
X—Cathead Channel—11 ft. 4 in. long —Left Hand
Y—Sheave Wheel
Z—Cathead Bearing

Q—Landing Panel Clamp—Front Half
R—Landing Panel Clamp—Back Half

Cage Parts

Letter	Description
CA—Cage Sill Angle—Sides
CB—Cage Sill Angle—Front and Back
CC—Cage Sill Conn. Plate—$\frac{1}{4}$ in. Plate
CD—Sill to Guide Conn. Plate—$\frac{3}{8}$ in. Plate
CE—Cage Guide Angle—7 ft. 5 in. long—Right Hand
CF—Cage Guide Angle—7 ft. 5 in. long—Left Hand
CG—Spacer Plate—$1\frac{1}{2}$ in. × $8\frac{1}{2}$ in.

Letter	Description
CH—Lift Bar Angle
CJ—Lift Bar Bracket
CJ1—Lift Bar Bracket Angle—Right Hand
CJ2—Lift Bar Bracket Angle—Left Hand
CK—Clevis—1 in. Diam.
CL—Cage to Guide Brace—$1\frac{1}{2}$ in. × 6 ft. $0\frac{1}{4}$ in.
CM—Wood Platform

CHAPTER XI

PAINTING AND REPAIRING STEEL STACKS

The painting and repairing of steel smokestacks are a perpetual headache to the maintenance men in large and small plants alike. In traveling about, one will observe numerous sheet-steel stacks on laundries, bakeries, small factories, etc., many of which are in a poor state of repair, due primarily to the fact that the plant maintenance men are unable to take care of them properly.

These men cannot be held entirely to blame for the poor condition of their stacks, for such maintenance is not in their usual line of duty. Most of these maintenance men have never been up on a stack in the course of their work and probably hesitate to go up unless the safest kind of rig is available. One look at the rusty steel cable hanging from the top of the stack usually convinces them that they were not born to be steeplejacks. When the stack was erected, a heavy steel hook was engaged over the top sheet and a steel pulley block attached to it, after which a galvanized steel cable (called a "gantline") was reeved through it. But years of exposure to the rain and to the flue gases have caused corrosion of the wire rope notwithstanding the galvanizing. It is customary to attach a simple boatswain's chair to the end of this gantline and to hoist a man to the top, carrying with him a new gantline hook and a set of manila-rope falls which are to be used in painting or repairing the stack. If the permanent gantline is not too badly corroded, he may get to the top safely; otherwise . . . ?

Not long ago, a painter on reaching the top of a stack in this conventional manner discovered that the hook on the pulley block on which his life depended was very badly pitted and corroded away. He managed to get down safely but soon began to work up a device to enable a man to reach the top of a small stack in safety. After considerable study and experimenting he developed the safety-type gantline hook shown in Fig. 1, which has proved very satisfactory. A steel bar is forged to the dimensions shown in the sketch, and the permanent gantline, which may be of a questionable strength, is fastened to the middle eye. A set of manila-rope falls that has just been inspected and found in safe condition is attached to the eye at the base of the hook, and the hook of the pulley block is moused to prevent accidental detachment. A tag line is then tied to the end eye, and the rig is hoisted by means of the gantline. When raised to the extreme position,

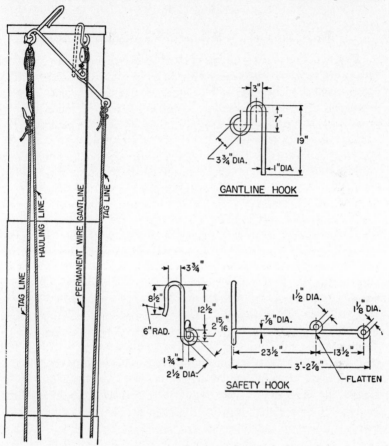

GANTLINE HOOK

SAFETY HOOK

Fig. 1. A special rig to provide a safe means of ascending a steel stack. (*Consolidated Edison Co. of New York, Inc.*)

the gantline is held stationary while a slight pull is taken on the tag line. This causes the bar to tilt and the hook end of it to pass over and engage the top of the stack.

The rope falls that were hoisted in a "block and block" position are pulled down by means of another tag line, and the boatswain's chair, preferably one of the safety type with a body belt to keep a man from falling out, is attached to the rope falls, and the painter hoists himself to the top in safety. Never does he trust his life to equipment except that which he has been able to inspect only a few minutes before. Upon completion of the job, the rig is unhooked by holding the gantline taut while pulling on the tag line. This detaches the safety hook from the top of the stack. The rig is then lowered and stored in a proper place until next required.

On intermediate-size stacks permanent steel ladders are usually provided. These ladders should be of standard dimensions, namely 18 in. wide, 12-in. rung spacing, and $6\frac{1}{2}$-in. toe room behind the ladder. Due to the necessary exposure to the weather and to the flue gases, the ladder should be of heavy construction, the rungs being not less than $\frac{3}{4}$ in. diameter and the stringers about $3 \times \frac{1}{2}$ in. On all long ladders, such as on stacks, it is desirable to provide basket guards or cages (Fig. 2) inside of which a man can climb in comparative safety, but on the intermediate-size stack a basket guard may appear all out of proportion to the size of the stack. In fact, it may look more like a basketed ladder with a stack attached to it. Consequently, basket guards are seldom used except on the very large stacks.

On these medium-size stacks and on those of larger size which are not provided with basket guards on the ladders, it may be found desirable to stretch a bronze or galvanized steel cable or wire strand up each side of the ladder from the very bottom to the very top of the stack, as shown in Fig. 3. These cables are alternately anchored to the ladder stringers about every 12 to 15 ft and are then pulled taut by means of turnbuckles. A life belt (Chap. XII, Fig. 3) is provided for each man and is equipped with two forged-steel (not malleable iron) snap hooks, as shown. When starting up the ladder the man snaps his two hooks onto the respective cables. After he has climbed about 15 ft, one of the hooks strikes the obstruction. He then detaches that hook and snaps it onto the cable above the anchorage, after which he continues to climb. Fifteen feet farther up the ladder his other hook strikes a similar obstruction and must be unhooked and hooked on again. This is repeated until he reaches the top of the ladder.

When disengaging one of the hooks, the man is still protected by the hook at the other side of the ladder. Consequently, there is never a time when he is not safeguarded by at least one hook. Should he feel weak or dizzy or tired, he can attach one of the hooks to the ladder rung for greater security while standing still. It is admitted that the use of such equipment is largely psychological, but never under any circumstances can he fall farther than 15 ft. Should he begin to fall, he will be held close to the ladder and will have a better chance of grasping it again. There is another advantage, and perhaps most important of all, in the use of this equipment on stack ladders: A man must stop and rest every 15 ft, even

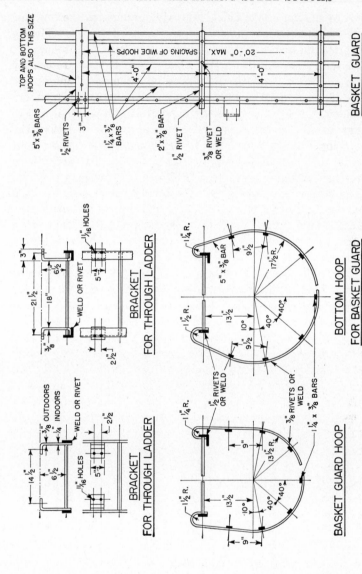

FIG. 2.   Details for safety basket guard used on tall steel stacks.   *(Consolidated Edison Co. of New York, Inc.)*

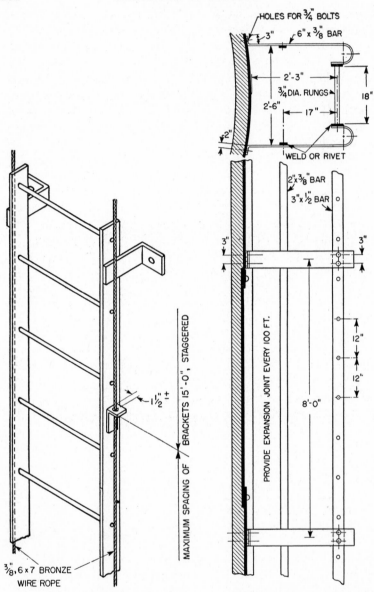

FIG. 3.   Safety cable installed beside ladder on steel stack on chimney. (*Consolidated Edison Co. of New York, Inc.*)

FIG. 4.   Details of another type of protection for a stack ladder. (*Consolidated Edison Co. of New York, Inc.*)

if for only a few seconds while reattaching his hooks, and he will arrive at the top in a much better physical condition.   Some riggers take pride in proclaiming that they can climb a 250-ft stack without once stopping to rest, but they do not tell how near they come to exhaustion or to having cramps in their arm or finger muscles, which might have fatal results for them.

When it is necessary to climb a long ladder, such as on a tall chimney which is not equipped with any safety device, the Safe-Hi ladder climber will provide complete safety for the man. A body belt worn by the man has two webbing straps of arm's length attached at opposite sides. These straps terminate in flat hand-shaped metal hooks with a latch which locks onto the rungs of the ladder (see Chap. XII, Fig. 4).

There are also similar safety belts which have a clamp that slides freely up or down a wire rope stretched up the center of the front face of the fixed ladder. Under normal conditions, due to its weight, the clamp hangs slightly below the man's belt. However, if he should start to fall his body would tend to drop faster than the clamp, due to its friction on the rope, and this causes the cam in the clamping device to jam on the rope. Hence, his free fall should be limited to not more than two or three feet.

On some large power-plant stacks the ladder is placed about 27 in. away from the stack wall, and the man climbs within this space. Thus, when tired he has merely to lean back against the stack and rest. A further improvement on ladders of this type is the installation of steel guard bars at each side of the ladder well (Fig. 4) to form an enclosure within which he climbs. With the bars located as shown in the sketch, it offers adequate protection for the man against being blown off the ladder by a high wind, yet it does permit him to climb out onto a scaffold if necessary.

Shown in Fig. 5 is a stack 21 ft 0 in. diameter at the top and 31 ft 0 in. at the boilerhouse roof with one continuous ladder over 300 ft high. Not only is this ladder equipped with a standard basket guard, but at 50-ft intervals are special seats formed by offsetting the bars of the basket guard. Steel plates are provided for the seat and the back rest, and the bars at the top of the recess in the ladder well are at such an angle as to reduce the hazard to a man's head or shoulder when he is about to start climbing the ladder again. While intended primarily as a seat, these landings are sometimes used to stand on while the men rest their feet. Also, these seats offer a safe and convenient means for a man climbing the ladder to pass another man coming down. This type of ladder and basket guard is considered the most desirable in use at present.

But merely getting to the balcony at the top of a large stack does not solve the entire problem. This balcony should be protected by a handrail at least 42 in. high and a 4-in. toeplate to prevent tools, etc., from being kicked off the edge. Some stacks have the top of the stack wall about $4\frac{1}{2}$ or 5 ft above the grating, which does not afford an opportunity for a man to stand on the platform and look down inside the stack. In such cases small checkered plate steps are provided at several places around the periphery of the stack at about 12 or 15 in. above the grating, so that a man can conveniently inspect the upper portion of the stack lining. Where such steps are provided, the handrail is made higher so as to be at least 42 in. above these steps. This may require the railing to be three or four members high, as these bars are usually spaced somewhat closer than on ordinary handrails. Where it is not

Fig. 5. This protected stack ladder is 304 ft high. Seats are provided every 50 ft for resting, or to enable men to pass each other on the ladder. (*Consolidated Edison Co. of New York, Inc.*)

feasible to increase the height of the railing, an eyebolt is placed at each side of each of the small steps just below the stack crown ring, and to these are attached the snap hooks of the body belt used in climbing stacks as mentioned above. The use of such a belt prevents him from falling down the outside or the inside of the stack.

In order to facilitate painting large power-plant stacks, a special articulated circular steel scaffold (Fig. 6) has been found very useful.

This scaffold was designed for use either on the outside or inside of stacks of about 22 ft diameter. It consists of 12 trapezoidal-shape platforms (part 10 in Fig. 7), each about 5 ft 6 in. inboard length, 6 ft 7 in. outboard length, and 2 ft 0 in. wide. The frame is constructed of $\frac{3}{16}$-in. steel plate, bent to form an L or angle shape with a 3-in. horizontal leg, which supports the floor boards, and a 4-in. vertical leg, which acts as a toeplate to prevent tools, etc., from rolling off the scaffold. These bent plates are mitered and welded together at the corners of the platform.

Fig. 6. With this special scaffold a stack 22 ft in diameter and 165 ft high can be painted in 8 hr. (*Consolidated Edison Co. of New York, Inc.*)

The floor boards (parts 1 and 2), which are the only wooden parts of the scaffold, are $1 \times 5\frac{1}{2}$ in., spaced $\frac{1}{2}$ in. apart and bolted to the platform frame.

Between the ends of the adjacent platforms are separators (part 9) made in the form of an inverted U. The horizontal web portion of the bent $\frac{1}{8}$-in. plate is 3 in. wide by 22 in. long and has vertical flanges at both sides 3 in. deep. Passing through the end toeplates of the adjacent platforms and through short lengths of tubing welded to the separator plate are two $\frac{3}{4}$-in. bolts 8 in. long, the nuts on which are left about 1 in. slack. Cotter pins are provided for security on these bolts. On the top of the bent separator plate is a hinged plate $3 \times \frac{3}{8} \times 12$ in. long, which is thrown back onto the separator while the scaffold is being hoisted and swung out to form an outrigger and stabilizer during lowering and working operations.

The scaffold is necessarily discontinued at the ladder, and in order to tie the platforms together into a continuous ring around

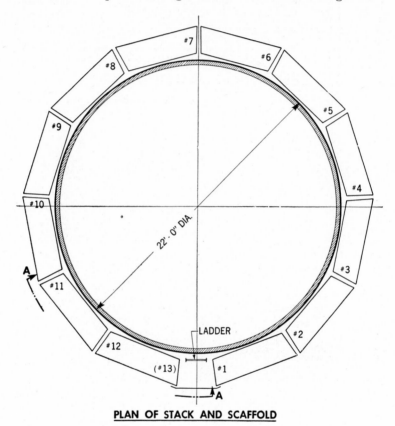

PLAN OF STACK AND SCAFFOLD

FIG. 7. Shop details for the special stack scaffold shown in Fig. 6. (*Consolidated Edison Co. of New York, Inc.*)

the stack a steel bar (part 6) or a $\frac{3}{4}$-in. rod is extended from the outboard corner of one platform to that of the other platform adjacent to the ladder.

At the 11 joints between platforms and at the two platform ends at the ladder, are standard Patent Scaffolding Company hand-operated scaffold winches (part 35) designed to accommodate about 250 ft of $\frac{5}{16}$-in. 6 $\times$ 19 galvanized-steel wire rope (part 27) for suspending the platform. These cables are provided with eye

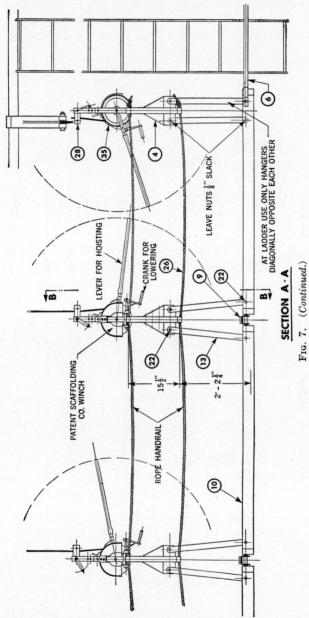

Fig. 7. *(Continued.)*

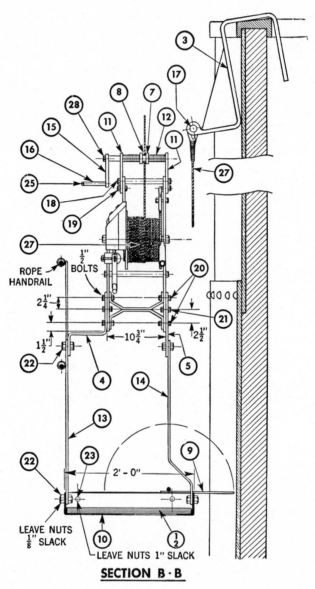

SECTION B·B

Fɪɢ. 7. (*Continued.*)

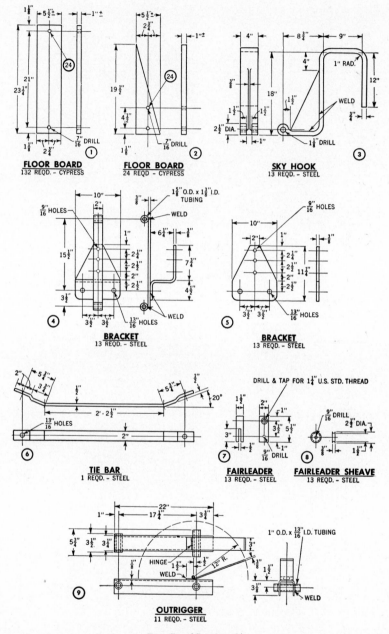

FIG. 7. *(Continued.)*

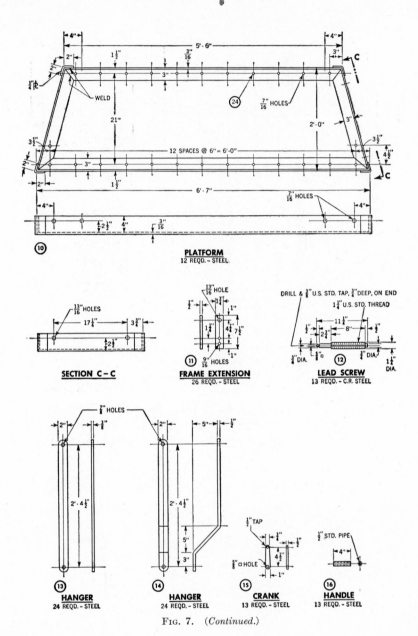

PLATFORM
12 REQD. ~ STEEL

SECTION C – C

FRAME EXTENSION
26 REQD. - STEEL

LEAD SCREW
13 REQD. - C.R. STEEL

HANGER
24 REQD. - STEEL

HANGER
24 REQD. - STEEL

CRANK
13 REQD. - STEEL

HANDLE
13 REQD. - STEEL

FIG. 7.   (*Continued.*)

BOLTS – STEEL

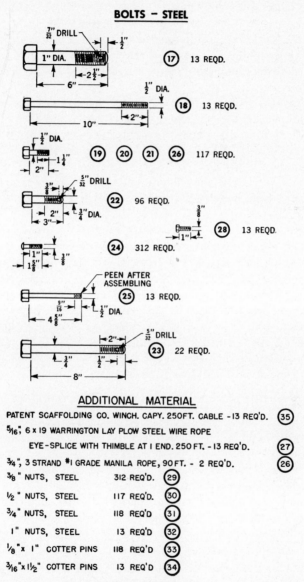

ADDITIONAL MATERIAL

PATENT SCAFFOLDING CO. WINCH. CAPY. 250FT. CABLE -13 REQ'D. (35)

5/16", 6 x 19 WARRINGTON LAY PLOW STEEL WIRE ROPE

    EYE-SPLICE WITH THIMBLE AT I END. 250FT. -13 REQ'D. (27)

3/4", 3 STRAND #I GRADE MANILA ROPE, 90FT. - 2 REQ'D. (26)

3/8 " NUTS, STEEL	312 REQ'D.	(29)
1/2 " NUTS, STEEL	117 REQ'D.	(30)
3/4 " NUTS, STEEL	118 REQ'D	(31)
I " NUTS, STEEL	13 REQ'D	(32)
1/8 "x I " COTTER PINS	118 REQ'D	(33)
3/16 "x 1 1/2 " COTTER PINS	13 REQ'D	(34)

FIG. 7. (*Continued.*)

splices and thimbles at the upper end for attaching to the "sky hooks" (part 3), which engage the top of the stack wall.

In raising the scaffold, the men "pump" the levers on the winches. In lowering, the main dog is held disengaged with one hand while the hand crank on the worm shaft is turned with the other. Never should the dog be wired or tied in the disengaged position.

The winch parts are all doubly protected against failure, there being duplicate dogs on the lowering motion and the main dog and the worm drive to hold when hoisting. Two of these devices would have to fail simultaneously to permit the winch to unwind accidentally by gravity.

To ensure winding the cable in even layers on the winch drum during the hoisting operation fair-leader (parts 7, 8, 11, 12, 15, 16, 25, and 28) is attached to the upper part of the winch frame, thereby making it possible to feed the rope on the drum as desired.

Special plate brackets (parts 4 and 5) are bolted to the lower end of the winch frame, and from those brackets are suspended four hanger bars (parts 13 and 14), which, in turn, are bolted to and support the corners of the adjacent scaffold platforms. These hangers do not lie exactly in the planes of the brackets or of the platforms, so in order to avoid the necessity of making them with very slight right- and left-hand twists, respectively, it was found more practicable to allow some play in the bolt holes and to leave about $\frac{1}{8}$-in. slack when placing the nuts on the bolts. Of course, all these bolts are provided with cotter pins.

At the ends of the scaffold near the stack ladder only two hangers are required, and these are attached to diagonally opposite holes in the two winch brackets so as to keep the winch frame in a vertical position and the drum shaft radially with relation to the stack.

With the platforms thus supported and tied together loosely, as previously described, the scaffold is sufficiently flexible to permit one winch to be raised a foot or more above the next. Although it is not essential to keep the scaffold absolutely level, it should be maintained reasonably so. To accomplish this when hoisting, the men should pump rather uniformly at the winches. Some gangs have found it practicable and entertaining as well to have one of their members sing loudly while they all keep time with their levers.

In order to pass from one platform to the next, it was necessary at the time the photograph (Fig. 6) was taken for a man to crawl under the winch. Since that time, however, the inboard hangers (part 14) and brackets (part 5) have been redesigned as shown in Fig. 7. This allows sufficient space for a man to pass between the winch and the wall of the stack by merely stepping over the offset lower ends of the hanger bars. The triangular openings between the platforms and the stack wall at the winches are small enough to prevent a man from falling through. When the outrigger is hinged into the "out" position, it forms a convenient step at this location.

Provision is made at each winch bracket (part 4) to support two lengths of ¾-in. No. 1 grade manila rope (part 26), which act as a handrail along the outboard edge of the scaffold.

The purpose of the tie bar (part 6) is to bind the entire scaffold into a unit and to have it completely encircle the stack in order to prevent the ends near the ladder from moving away from the stack wall. Also, if by some stretch of the imagination it could be assumed that 12 out of the 13 suspension cables had failed, the remaining cable would be expected to cause the scaffold as a unit to tilt and bind on the stack, thus preventing it from dropping farther.

This same scaffold is designed for use on the inside of the stack, in which case only nine platforms are used. When erected inside the stack the winch brackets are reversed in relation to the platform so that the rope railing will be on the inboard (unprotected) side of the scaffold. The actual erection of the scaffold (exclusive of any false work that may be required if the boilerhouse roof is not flat) usually can be accomplished in one day. On the second day 13 men (they are not superstitious even on a job of this kind) place their paint pots and tools on the scaffold and pump it to the top of the stack, just below the balcony. They then paint their way down to the boilerhouse roof, using ordinary brushes and making the entire operation in one 8-hr day on a stack 22 ft 0 in. diameter by 165 ft high above the roof. That portion of the stack above the balcony is usually painted on the preceding day. This type of scaffold has been in use about thirty-five years and has met with the unanimous approval of the painters and construction men.

Repairing brick chimneys is not usually considered within the field of industrial rigging, but rigging has to be done by the steeplejack, especially if no fixed steel ladder has been provided. Hence, it may be in order to show how he reaches the top of the chimney and what type of scaffold he uses. (See Fig. 8.)

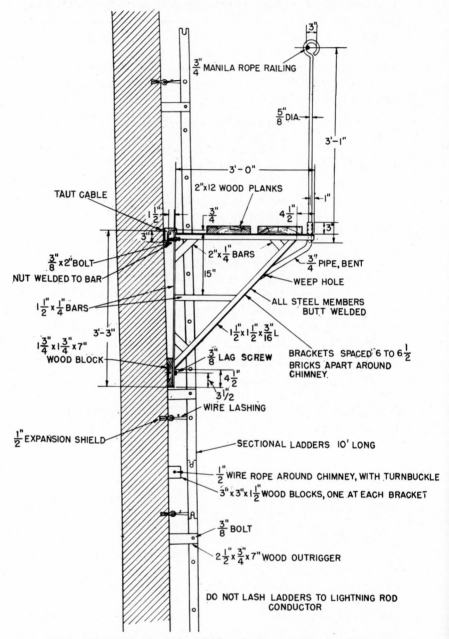

FIG. 8.   Portable ladder and scaffold used when pointing up a brick chimney.

A 10-ft sectional ladder with small outriggers attached is stood against the chimney, and the bottom and top rungs are lashed to expansion bolts inserted into the masonry. The next section of the ladder is carried up and dovetailed into the top of the first ladder, and the bottom rung anchored to the chimney. It is then necessary to climb up this partially secured ladder, drill the masonry and insert an anchor bolt, and then lash the top rung of

FIG. 9. Head, back, shoulder, and toe protection recommended for men working inside or outside at base of chimney being repaired.

this ladder section. The scaffold brackets are hoisted and hooked onto a tight cable which has been placed around the chimney. This kind of a job is necessarily dangerous, to say the least.

When repairs are made to the brick lining of a stack, regardless of the precautions taken, chips of brick will frequently fall to the base of the stack, presenting a serious hazard to the men who are at work there. In order to reduce this hazard to a minimum, the men are always provided with safety hats. On some jobs the men have, in addition, been furnished with protection for their shoulders and their backs when stooping. Figure 9 shows pads of Celotex covered by lightweight canvas and made into the form of a vest.

CHAPTER XII

LIFE BELTS, BOATSWAIN'S CHAIRS, AND LIFE NETS

There are a number of appliances or devices that serve a very useful purpose on rigging jobs, especially as regards personal safety. Among these are life belts, life lines, boatswain's chairs, parachute harness, and life nets.

Life Belts. Different designs for life belts are available for various special uses. Only webbing-type belts should be used, as leather deteriorates very rapidly and may fail when it is called upon to perform in an emergency. Leather is also very difficult to inspect visually. Some belts are made with only a body belt, some with a body belt and a wider cushion of webbing to reduce "cutting" of the body when the fall is suddenly arrested, and some with shoulder straps to take the weight of the belt. An objection to the last type is that the shoulder straps are frequently made so short that the body belt, instead of being around the waist or soft part of the body, is up onto the lower ribs, which may be fractured by the sudden strain on the belt. Also, this type is not too comfortable to wear, particularly in warm weather.

Figure 1 shows a simple life belt that has proved satisfactory in service. The buckle is of ample strength and is of a type that does not require puncturing the webbing at the buckle. It is a very comfortable belt to wear. Depending upon its use, several different types of tail lines may be used. If the belt is to be used to rescue a man entering a hazardous location or area, such as a coal bunker or a smoke- or gas-filled room, a long rope is used.

If this belt is used on a scaffold in conjunction with a hanging life line, a tail line 5 or 6 ft long is used, and this short rope is attached to the life line by means of a rolling hitch (see Chap. II, Fig. 49), or by means of a special friction device (Fig. 2). If the belt is worn by a man at work on the box scaffold (see Chap. IX), the tail line shown in Fig. 1 is used, the snap hook being attached to the trolley cable on the scaffold railing. This tail line can be made with or without the rubber shock cord, but preferably with it, as in the event of a man falling, the rubber can stretch about

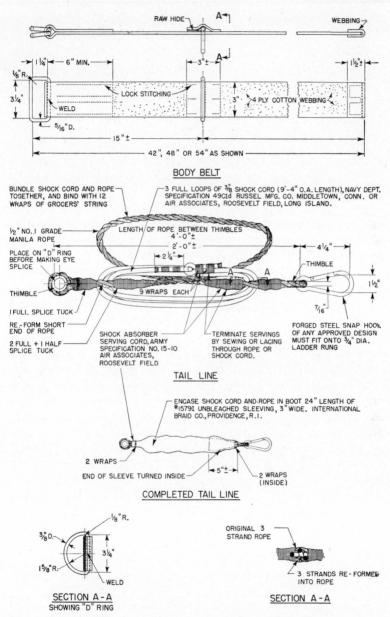

RAW HIDE

WEBBING

BODY BELT

TAIL LINE

COMPLETED TAIL LINE

SECTION A-A
SHOWING "D" RING

SECTION A-A

Fig. 1. A simple and comfortable life line belt that affords safety to men at work at elevated locations. (*Consolidated Edison Co. of New York, Inc.*)

Notes: (1) Hardware shall be of mild steel, cadmium-plated. (2) Butt welds shall be ground smooth. (3) Cotton webbing shall weigh not less than 16 lb per

18 in. and thereby ease the shock of deceleration.   If the shock
cord is not used, of course the rope will be shorter in order to
retain the 2 ft 0 in. over-all length.   The shock cord can also be
incorporated in the tail line used with the hanging life line.

A life line should be strong enough to arrest the fall of a man
but need not be excessively heavy or cumbersome.   Of course
only new or relatively new rope should be used and should be
secured to some substantial object overhead.   A new ½-in.
manila rope has a breaking strength of at least 2,650 lb.   Any
impact that will develop a force of 2,650 lb will undoubtedly
break a man's back and kill him; therefore there is no advantage
to be gained by using, say, a 1-in. rope.   A ½-in. tail line and a
¾-in. hanging life line are frequently used.

Fɪɢ. 2.   Sky-Genie snubber for attaching the tail line of a man's life belt to a
hanging life line.   (*Atlas Safety Equipment Co.*)

For climbing long ladders such as on chimneys or gas holders,
the belt shown in Fig. 3 is very useful.   The two snap hooks are
attached to the two cables shown in Fig. 3, Chap. XI.   As the
man climbs the ladder, the hooks slide up the ropes.   When an
obstruction is encountered, he stops, unhooks the hook, and
again attaches it above the anchorage.   In this way one hook is
always attached.   If he has to work from the ladder, the hooks
may be snapped onto the rungs.

For climbing and/or working on long or high fixed ladders, the
Safe-Hi ladder belt assures completely safety, even for a novice.
This device consists of a webbing body belt with two arm's-length
straps, one at each side of the man's body (Fig. 4).   These straps

---

100 linear ft.   (4) Lock stitching shall be as shown.   At the buckle and at the
"D" ring stitching shall not run transverse to belt.   (5) Stitching shall be done
with five-cord linen thread, hot-waxed, and having a tensile strength not less than
30 lb.   Number of stitches shall be not less than five nor more than six per inch.
(6) Shock cord shall be not more than 90 days old when belt is fabricated.

Tracers in braid of shock cord to indicate its age: Two yellow threads together—1964; two green threads together—1965; two red threads together—1966;
two blue threads together—1967, and repeat; one red thread alone—January,
February, March; one blue thread alone—April, May, June; one green thread
alone—July, August, September; one yellow thread alone—October, November,
December.

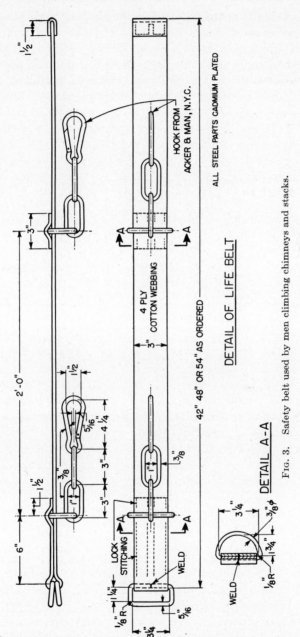

DETAIL OF LIFE BELT

DETAIL A-A

Fig. 3.   Safety belt used by men climbing chimneys and stacks.

terminate in flat stainless-steel hooks which are held by inserting the hands under their small straps.

These hooks protect the hands from abrasion due to rusty rungs, and a man need not wear gloves when climbing the ladder. Each time the hand (with the hook) is placed on a rung, a latch automatically locks it in place.   Under normal conditions, when

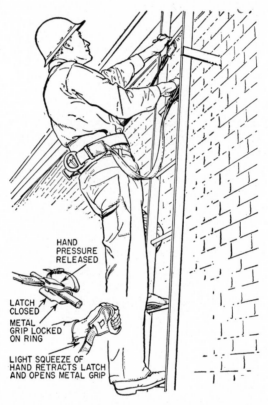

HAND
PRESSURE
RELEASED

LATCH
CLOSED
METAL
GRIP LOCKED
ON RING

LIGHT SQUEEZE OF
HAND RETRACTS LATCH
AND OPENS METAL GRIP

FIG. 4.   Safe-Hi belt makes climbing any high ladder safe.   (*Rose Mfg. Co.*)

letting go of a rung the man relaxes his grip, and he would do likewise unconsciously if he should collapse on the ladder, such as from a heart attack or fright.   But in this case, the hooks remain attached to the rungs.

To let go of the rung he has to squeeze the hook or grip with his fingers, thus causing the palm of his hand to press against a lever which retracts the latch.   Should the man need to work from the ladder he merely removes his hands from the grips, which remain locked to the rungs.

Boatswain's Chairs. For performing small jobs at otherwise inaccessible elevated locations, a boatswain's chair is frequently used. The original type consisted of a flat board with one or two holes near each end (Fig. 5), through which $\frac{3}{4}$- or 1-in. manila rope was threaded. But in using this rig, there is always the possibility of a man's falling out owing to sickness or inattention.

Usually the boatswain's chair is supported by a three-part rope tackle (one single and one double block), the man raising and lowering himself, or being assisted by a man on the ground below. Infrequently, as in tree work, a man may get to a limb by other means and then desire to lower himself a few feet at a time until he eventually reaches the ground. For such a job the rope is passed over a large limb, one end of the rope being attached to the boatswain's chair and the other end, with sufficient slack, remaining on the ground.

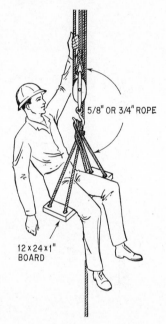

5/8" OR 3/4" ROPE

12 x 24 x 1" BOARD

FIG. 5. The common boatswain's chair.

The tail end of the rope, above the bowline knot, is attached to the hanging part of the rope by means of a rolling hitch. Then, to lower himself the man grasps the hanging line above the hitch with his left hand, and with his right hand he slides the hitch down the rope, thus lowering himself. As a safety precaution, a figure-eight knot should be tied at the end of the tail line to prevent accidental untying.

A safety-type boatswain's chair, such as the one shown in Fig. 6, is recommended for use. This has a comfortable "harvester" seat and a body life belt. In order to prevent restriction of blood circulation in the man's feet (feet going "to sleep"), stirrups are provided in which to rest the feet. Also, to reduce the discomfort to his knees when working against a wall for a length of time, casters are provided to keep his knees away from the wall.

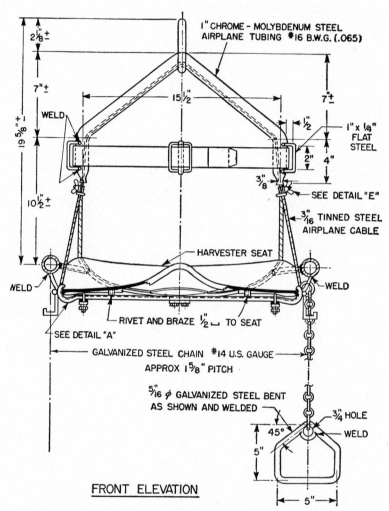

FIG. 6. This "safety"-type boatswain's chair is not only safer but also more comfortable for the man's seat, knees, and feet. (*Consolidated Edison Co. of New York, Inc.*)

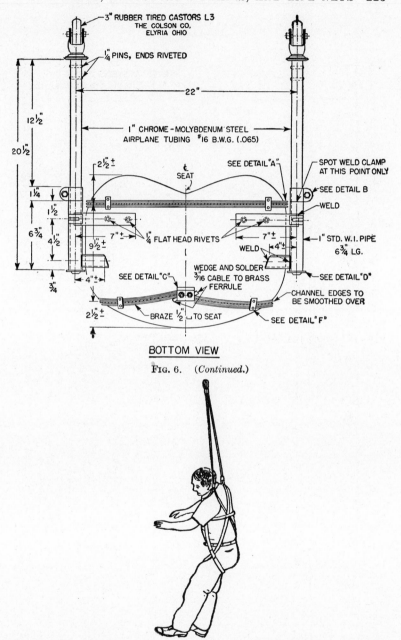

BOTTOM VIEW

FIG. 6. (*Continued.*)

FIG. 7. The parachute harness can be used in place of a boatswain's chair when clearance is restricted.

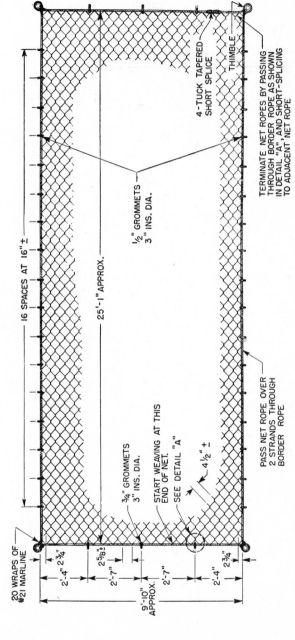

FIG. 8.   A life net similar to those used by circus acrobats is most resilient and will stop a man less abruptly.   (*Consolidated Edison Co. of New York, Inc.*)

NOTES: (1) All ropes shall be of three-strand manila bolt rope or nylon.   (2) All short splices shall have three full tucks (four for nylon) and then be neatly tapered.   (3) Grommets on ends of net are for lacing to supporting structure by means of ¾-in. manila rope.   (4) Grommets on sides of net are for lacing to adjacent net by threading a ½-in. manila rope alternately through grommets on the two nets, and securing to the supporting structure.   (5) Dimensions shown (except rope sizes) are approximate only.   (6) Nets should be neatly folded when not in use and hung up on wooden pegs in a cool, dry location with free circulation of air.

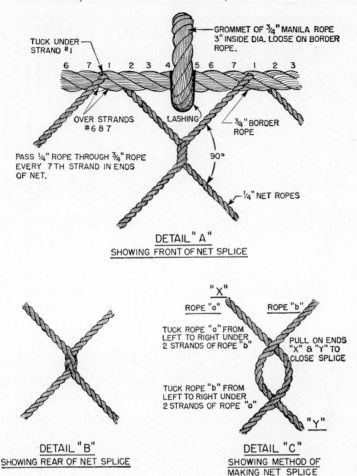

TUCK UNDER STRAND #1

GROMMET OF ¾" MANILA ROPE 3" INSIDE DIA. LOOSE ON BORDER ROPE.

6 7 1 2 3 4 5 6 7 1 2 3

OVER STRANDS #6 8 7

LASHING

¾" BORDER ROPE

PASS ¼" ROPE THROUGH ¾" ROPE EVERY 7TH STRAND IN ENDS OF NET.

90°

¼" NET ROPES

DETAIL "A"
SHOWING FRONT OF NET SPLICE

DETAIL "B"
SHOWING REAR OF NET SPLICE

"X"

ROPE "a"   ROPE "b"

TUCK ROPE "a" FROM LEFT TO RIGHT UNDER 2 STRANDS OF ROPE "b"

PULL ON ENDS "X" & "Y" TO CLOSE SPLICE

TUCK ROPE "b" FROM LEFT TO RIGHT UNDER 2 STRANDS OF ROPE "a"

"Y"

DETAIL "C"
SHOWING METHOD OF MAKING NET SPLICE

FIG. 8. (*Continued.*)

The safety latch on the pulley block (Chap. IX) also adds to the man's security.

Parachute Harness. Occasionally it is necessary to hoist or lower a man through a maze of piping or other obstruction or through a man-hole or small hatch, in which a boatswain's chair would be too large or too cumbersome. In this case a webbing parachute harness (Fig. 7) is sometimes used. This fits snugly around the man's body and takes practically no more room than his body. The parachute harness, however, is not so comfortable as the safety type boatswain's chair for long operations.

If a parachute harness is not available, it may be necessary to resort to a body sling made from a bowline-on-a-bight knot (Chapter II, Fig. 23).

Life Nets. On some high jobs, such as erecting or painting steel structures where it is impossible to wear life belts, it may be practicable to provide life nets below the point where the men are at work. A life net, to be of value, should be placed as near as possible to the level where the men are working; in other words, their falling distance should be reduced to a minimum. Also, keep in mind as in the case of life belts that it is not the fall but rather the sudden stop which causes injury to the man. Therefore the life net should be made as springy as possible. Nylon net ropes are recommended.

In falling, a man (or any body) develops energy in foot-pounds equal to his weight in pounds multiplied by the distance he falls in feet. This energy must be absorbed when his fall is stopped. Therefore, the force of impact is equal to the energy divided by the distance in feet in which he is stopped. For instance a 180-lb man falls 25 ft and develops 4,500 ft-lb of energy. If he is stopped in 1 ft, the force is 4,500 lb. If stopped in 2 ft, it is 2,250 lb. In 5 ft the impact is 900 lb, etc. To this figure must be added the weight of the man, for his weight would have to be supported even though he did not fall.

Some nets are made with mesh ropes parallel to the border ropes, thus forming squares. A man falling into the net will be stopped quite suddenly, as there will be relatively little stretch to the mesh ropes.

The net shown in Fig. 8 follows quite closely the design used by Ringling Brothers–Barnum and Bailey Circus in the nets for its acrobats. The mesh ropes, running at a 45-deg angle, form diamonds rather than squares. Thus, when a man lands in the net, the diamonds are distorted and the shock is absorbed by practically all the ropes. The net sags considerably under the load of a man, and consequently the shock to the man and the strain on the net are at a minimum. The net can be made by splicing the mesh ropes as shown in detail in Fig. 46 in Chap. II or by making the sheet bend (Fig. 44 in Chap. II).

Throwing a Rope. Occasionally it is necessary for a rigger to throw a rope over a beam or other support in order to suspend a set of rope falls or to enable him to climb up to that elevation. In such a case, the instructions given by the National Park Service in their booklet "Rope Knots and Climbing," Tree Preservation Bulletin No. 7 for foresters, can be used equally well by riggers. For general use a $\frac{1}{2}$-in. manila rope is recommended.

In order to throw a rope over an elevated beam, pipe, or tree branch, use a length of rope which has just been inspected. Make a throwing knot which will be heavy enough to be thrown to the required height, but which can uncoil itself as it is paid out, allowing the end to return to the floor or ground.

Holding the end of the rope in the left hand (Fig. 9) (a left-handed man should reverse the following directions) make 8 or 10 loops about 8 to 10 in. long, then wrap these loops with about

Fig. 9. Making a throwing knot, first position.

8 turns, and pull a bight of the rope through the upper portion of the coil (Fig. 10). The knot should be kept tight and compact so that there will be no loose loops which might catch on projections.

For throwing the rope the bight held in the left hand (Fig. 10) is adjusted so that the coil just about reaches the ground when the arm is straight. Then five or six loops of the same length are made. The knot and one or two loops are held in the right hand, and the balance of the loops are held in the left hand as shown in Fig. 11.

When throwing the rope, the rigger should stand facing the

beam, pipe, or tree branch at a point that will allow a free throw of the throwing knot over the support. With the right knee bent and the weight carried on the right leg, the arms should be swung forward and backward together to get momentum. After two or three swings, the right hand should release the throwing knot while the rope held in the left hand should be released loop by loop. If the throwing knot has been aimed properly, it will go over the support and fall on the other side. The knot should release itself and the rope end fall to the ground. If the end fails to come within reach, it will be necessary to throw running coils

Fig. 10.   Making a throwing knot, second position.

up the rope to whip it over the support. If the support is too high to reach with a $\frac{1}{2}$-in. rope, a smaller rope tied to a weight may be used, and this rope used to pull up the $\frac{1}{2}$-in. rope.

Climbing a Pole or Tree by Using a Rope. After a rope has been thrown over the support, the two ends should be brought together and the double rope held in the hands (Fig. 12) as high as possible above the head. Then, with the weight of the body supported by the hands, the legs are quickly lifted as high as possible and wrapped around the pole or tree (see Fig. 13). Still holding tightly with the legs, climb the rope hand over hand. When the position shown in Fig. 14 is reached, the legs are

FIG. 11. Making a throwing knot, ready to throw.

FIG. 12. Ready to climb.

FIG. 13. Legs up.

released and again lifted as high as possible to a new position on the pole or tree.

In order to climb onto the supporting beam or tree branch, the right leg is swung up and hooked over the support which is then grasped with both hands, while the left leg is used as a brace against the pole.   The body is then swung up onto the supporting member.

<table>
<tr><td>Fɪɢ. 14.   Hand over hand.</td><td>Fɪɢ. 15.   Footlock,
first position.</td></tr>
</table>

If necessary to climb higher by means of a rope, one end of the rope is re-coiled into a throwing knot and thrown over a higher support.

Climbing by Means of the Footlock.   It is frequently necessary to climb a rope when the pole or tree trunk cannot be used to assist.   Never attempt to climb a rope hand over hand without using the footlock.   Experienced climbers may have their own preferred form of footlock, but the National Park Service footlock is suggested as an acceptable and proved standard.

With the double rope hanging free, grasp the two ropes as high

as possible, and raise the body with the arms. Then the right
knee should be raised with the ropes passing outside the right
thigh, back of the right leg, over the right instep, and over the
left instep as shown in Fig. 15. Then the left foot is raised and
placed on the right instep, locking the ropes between the left arch

FIG. 16. Foot-
lock, second posi-
tion.

FIG. 17. Footlock, third
position.

and the right instep (see Fig. 16). In this position the entire
weight of the climber should be borne by the feet. Then the
arms are raised (Fig. 17) to get a new grip on the rope, the footlock is released, and the legs raised and locked as before.

This process should be repeated as often as is necessary. When
the support is reached, one leg is thrown over it and the body

swung over and onto it.   The rope is then thrown over a higher
support, if necessary, and climbing is resumed.   Cramps and
strains should be avoided when using the footlock, as well as in
other types of climbing.   The climber should stop and rest if he
feels tired.   The careful climber should not use the footlock
method for a climb of over 20 feet.

For climbing a flag pole (Fig. 18) a short rope is placed around
the pole in the form of a clove hitch (Chap. II, Figs. 37 and 68),
with the ends of nearly equal length.   The longer end is placed

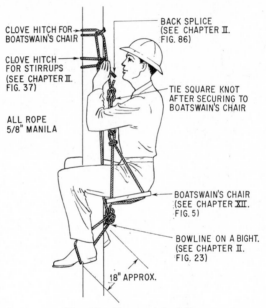

CLOVE HITCH FOR→
BOATSWAIN'S CHAIR

CLOVE HITCH→
FOR STIRRUPS
(SEE CHAPTER II.
FIG. 37)

ALL ROPE
5/8" MANILA

BACK SPLICE
(SEE CHAPTER II.
FIG. 86)

TIE SQUARE KNOT
AFTER SECURING TO
BOATSWAIN'S CHAIR

BOATSWAIN'S CHAIR
(SEE CHAPTER XII.
FIG. 5)

BOWLINE ON A BIGHT.
(SEE CHAPTER II.
FIG. 23)

18" APPROX.

FIG. 18.   Boatswain's chair.

through the eye of the spreader rope of a boatswain's chair, and
tied back onto itself by means of a square knot.   The hitch should
be about three feet above the seat.

The end of another short rope is tied into a bowline-on-a-bight
(Chap. II, Fig. 23), the loops being about 15 to 18 in. long.   The
other end of this rope, about four feet above the bottom of the
loops, is likewise attached to the pole below the other hitch by a
similar clove hitch.   The short end of this rope is made into a
back splice (Chap. II, Fig. 86) to prevent accidentally slipping
out, since the hitch has to be loosened when descending from the
pole.

The rigger gets into the boatswain's chair, suspended from the lower end of the pole. He then bends his knees, and slides the hitch of the foot rope up nearly to the seat hitch. Then he stands up in the bights of the foot rope, and slides the hitch for the seat up the pole. This is repeated, alternately moving the hitches up (or down) the pole. Keep in mind that when descending the pole, the pole diameter keeps increasing. This means loosening the hitches as required.

For safely lowering a person (whether active, injured, unconscious, or dead) from a precarious position high above the ground, such as a fireman trapped on the roof of a burning building, or a Boy Scout marooned on a rock ledge of a mountain cliff, the friction device shown in Fig. 2 may be very handy. It consists of a $\frac{3}{8}$-in. braided nylon rope (having a safe working strength of about 1,000 lb) which is secured at its upper end to a suitable anchorage, and allowed to hang vertically or nearly so. The rope makes several turns helically around the aluminum core of the device and thus acts as a friction snubber. To the eye at the bottom of this device is attached a life belt, parachute harness, boatswain's chair, basket stretcher (Fig. 4, Chapter XXII), or a body bag, as the need may be.

When the weight is suspended from the snubber it will be lowered at a controlled speed to the ground. The man being lowered may grasp the rope below the snubber between the thumb and forefinger, or someone on the ground may very gently pull on the hanging rope, and the descent will be slowed or stopped, as may be desired.

CHAPTER XIII

PORTABLE LADDERS

Many industrial organizations that use a large number of wood ladders have an inspector carefully examine each ladder when received in order to make certain that it meets the requirements of the purchase specifications.

Most specifications are based upon the data contained in the American Standard Safety Code for the Construction, Care and Use of Ladders, which is published by the ASA, or else they refer to this code. Therefore, the first duty of the inspector is to check all the dimensions of the ladder, including the size of the side rails and rungs, to assure himself that they are not less than the minimum allowed by the code (Table I). Well-built ladders usually have their side rails larger than the code dimensions. All wood parts, of course, should be free from splinters. Next, he should examine the hardware to determine if it is of ample strength. Malleable iron and cast iron should be avoided, if possible, for parts subject to bending or tensile stress. Rungs should be tightly fitted into the side rails and secured against turning by the use of finishing nails driven through the side rails and into the tenons, yet they should not be so tight as to split the side rails.

Special or "trick" ladders should not be used. These ladders, which can quickly be converted into an extension ladder or stepladder, may possibly be acceptable for household use, but they have no place in industry.

On stepladders the spreader bars should be so designed as not to present a serious finger-pinching hazard. Stepladders should not be able to fold up accidentally if pushed along the floor. Extension ladders should have the guide brackets long enough to engage the full width of the side rails on the other section of the ladder. Also, the locks should be of proper design, and the rope and pulley of ample strength. Near the lower end of the upper section of extension ladders a rung is often omitted at the location of the locks. A special offset rung may be necessary, but a rung of some description is essential, due to the fact that the sections of an extension ladder are frequently separated and used independently.

TABLE I. MINIMUM DIMENSIONS OF WOOD LADDERS

(Sitka spruce, Douglas fir, Southern yellow pine.)

Single Ladders*

Nominal length, ft	Width at base, inside, in.	Size of side rails, in.
up to 10	$11\frac{1}{2}$	$1\frac{1}{8} \times 2\frac{1}{2}$
12	$11\frac{3}{4}$	$1\frac{1}{8} \times 2\frac{1}{2}$
14	12	$1\frac{1}{8} \times 2\frac{1}{2}$
16	$12\frac{1}{4}$	$1\frac{1}{8} \times 2\frac{1}{2}$
18	$12\frac{1}{2}$	$1\frac{1}{4} \times 2\frac{3}{4}$
20	$12\frac{3}{4}$	$1\frac{1}{4} \times 2\frac{3}{4}$
24	$13\frac{1}{4}$	$1\frac{1}{4} \times 3$
26	$13\frac{1}{2}$	$1\frac{1}{4} \times 3$
28	$13\frac{3}{4}$	$1\frac{1}{4} \times 3$
30	14	$1\frac{1}{4} \times 3$

Extension ladders†

Nominal length, ft	Width at base, inside, in.	Size of side rails		
		Sitka spruce, in.	Douglas fir; Yellow pine, in.	Overlap, ft
up to 24	$14\frac{1}{2}$	$1\frac{5}{16} \times 2\frac{1}{2}$	$1\frac{5}{16} \times 2\frac{1}{2}$	3
26	$14\frac{1}{2}$	$1\frac{5}{16} \times 2\frac{3}{4}$	$1\frac{5}{16} \times 2\frac{1}{2}$	3
28	$14\frac{1}{2}$	$1\frac{5}{16} \times 2\frac{3}{4}$	$1\frac{5}{16} \times 2\frac{1}{2}$	3
30	16	$1\frac{5}{16} \times 2\frac{3}{4}$	$1\frac{5}{16} \times 2\frac{1}{2}$	3
36	16	$1\frac{5}{16} \times 3$	$1\frac{5}{16} \times 2\frac{3}{4}$	3
40	16	$1\frac{3}{8} \times 3$	$1\frac{5}{16} \times 2\frac{3}{4}$	4
44	18	$1\frac{3}{8} \times 3\frac{1}{4}$	$1\frac{5}{16} \times 3$	4
48	18	$1\frac{1}{2} \times 3\frac{1}{2}$		4
60	18	$1\frac{5}{8} \times 4$		5

If the dimensions of the material are increased the allowable angle of cross grain will be slightly greater.

All material should be seasoned and dressed, free from splinters and sharp edges, free from shake, wane, compression wood, compression failures, decay and low density.

Side rails: Cross grain should be not greater than $1:12$. No knots allowed on narrow faces. On the wide faces knots are permitted if not larger than $\frac{1}{2}$ in., if not closer than $\frac{1}{2}$ in. to the edge, and if not less than 3 ft. apart. Checks limited to $6 \times \frac{1}{2}$ in. Pitch pockets limited to $\frac{1}{8} \times 2 \times \frac{1}{2}$ in. and not less than 3 ft apart.

Rungs: not more than 12 in. apart. Not less than $1\frac{1}{8}$ in. diameter with tenons $\frac{7}{8} \times 1\frac{1}{8}$ in. long for ladders up to 36 ft long; $1\frac{1}{4}$-in. rungs for longer ladders. Cross grain limited to $1:15$. Knots limited to $\frac{1}{8}$ in.

* Data taken from American Standard Safety Code for Wood Ladders, ASA A14.1, 1959.
† American Standards Association.

Finally, the inspector should make a thorough examination of the wood itself. This is one part of the inspection which is usually omitted in part or in its entirety. To many purchasers and, strange as it may seem, even to some ladder manufacturers, wood is wood and nothing more. The data on wood contained in this chapter will therefore be devoted largely to the inspection of the material used in straight or extension ladders, these being the types which are responsible for most accidents.   For further data on the inspection of wood, refer to Chap. VI.   Most ladders of these types are constructed with spruce side rails (preferably Sitka spruce) and with oak or hickory rungs.

Oak and hickory are readily recognized, but spruce may sometimes be confused with other species.   Therefore, it may not be amiss to include a few words concerning the identification of spruce.   Spruce is nearly white, with no definite difference in color between heartwood and sapwood, except that Sitka spruce does have heartwood with a slight reddish tinge.   Spruce averages about 15 to 18 annual rings per inch as measured radially on the ends of the side rails and weighs about 28 lb per cu ft.   The summerwood of the annual rings is distinct, but not horny.   (For additional information on defects in wood, see Chap. VI.)

Resin ducts appear as white specks in the summerwood on the cross section and as faint pin scratches on the longitudinal faces. Pitch pockets are frequently observed, and resin may ooze from newly cut surfaces if exposed to warm atmosphere.   Spruce is distinguished from white pine by having fewer and smaller resin ducts than pine and also by the lack of distinct heartwood in all species of spruce except Sitka.   It is distinguished from fir by the presence of its resin ducts.   Sitka spruce usually has a sheen to the longitudinal surfaces, and the flat-grain face is generally "dimpled" or "pocked," thus distinguishing it from Douglas fir.   Sitka spruce comes from larger trees than other species of spruce, and its rings are therefore usually of longer radius.

After the inspector has satisfied himself that the proper kind of wood has been used, he should first check up on the direction of the annual rings as observed on the ends of the side rails.   These rings should preferably be approximately parallel to the narrow faces, thus producing edge grain on the wide faces and flat grain on the narrow faces of the rails.   With the rings extending in this direction the ladder will be somewhat stronger than if they ran in the other direction, but it will not be quite so stiff.   (A springy ladder is unpleasant to work upon, but it has greater resistance to sudden or impact loads.)   Also, the harmful effect of cross grain, if it exists, is reduced, and spiral grain is more readily detected by noting the tiny splits at the nails used to hold the rungs from turning.

Next, the presence of cross grain should be investigated.   Cross grain may be subdivided into "diagonal" and "spiral" types, and both either individually or collectively should not exceed an angle of 1 : 15 with the edge of the rail. Diagonal grain is observed on the edge grain face (wide face in this instance) and is indicated by the direction of the annual ring markings.

Dip grain, a local deviation from straight grain caused by knots, tree wounds, etc., which may or may not be present in the side rail, should also be given careful study. Such defects are most harmful if occurring in the middle of the length of a ladder. This also means in the top half of the lower section and in the bottom half of the upper section of an extension ladder. It is the angle made by the dip grain at the edge of the side rail that determines its importance, and this angle should not exceed 1:15. Farther back from the edge a slightly greater angle is allowable. Dip grain, of course, should not be permitted near rung holes.

Shakes are not permitted in ladder stock.

In edge-grain ladder stock, as was previously recommended, checks will seldom be found but if present should not exceed 6 in. in length or $\frac{1}{2}$ in. in depth.

Pitch pockets are not objectionable if they are not larger than 2 in. long, $\frac{1}{8}$ in. wide or $\frac{1}{2}$ in. deep, or if they are not closer than 3 ft apart.

According to the ladder code, knots are not permitted on the narrow faces. In edge-grain material, such as has just been recommended, this is the only place where knots could occur. Therefore, all knots are prohibited.

Only well-seasoned material should be accepted, as green wood will sooner or later cause the joints to loosen and the ladder will become rickety. Warped material is also unacceptable.

Some inspectors insist upon testing all ladders as a final requirement before acceptance. Such load testing is no longer recommended in the ladder code, and it is frowned upon also by authorities on wood, as there is far greater probability of producing a compression failure during the test than of discovering one that already exists. Consequently, all testing should be eliminated from the inspection routine.

Now to investigate the rungs. Hand-split rungs are preferred, as they are naturally straight grained. If the rungs have been turned, they should be checked for cross grain at two points 90 deg apart, the limit of the cross grain being 1 : 15. Knots over $\frac{1}{8}$ in. diameter, checks, and other defects are all prohibited. Rungs can be tested for cracks by striking with a light hammer.

In special ladders that are constructed with side rails considerably heavier than required by the ladder code, certain defects such as knots, checks, cross grain, etc., may be tolerated at the discretion of the inspector even though slightly exceeding the figures given previously if the strength of the ladder is otherwise considerably above the minimum strength. To enable the inspector quickly to check the strength of a ladder, the accompanying chart (Fig. 1) has been prepared. It is based upon an average-weight man standing on the middle rung of a ladder inclined at the recommended angle (the base of the ladder at a distance from the wall equal to one-fourth the length of the ladder). The net area after deducting rung holes has been used, and the unit stress developed should not exceed 1,600 lb per sq in.

In order to enable the inspector to make a thorough inspection, the ladders as received from the manufacturer should be of untreated wood; that is, no paint, varnish, linseed oil, or other surface protection should be applied before the inspection. (Unscrupulous manufacturers have been known to paint wood to conceal otherwise visible defects.)

The Forest Products Laboratory advises that on objects fabricated from wood, including ladders, which are coated with ordinary paint, varnish, or similar materials moisture may enter the wood at the uncoated joints such as rung tenons and rung holes. This moisture is then prevented from leaving the wood because of the more or less impervious coating. Hence, they advise that "in some kinds of wood construction involving such joints, the application of protection coatings may favor rather than retard decay."

There are available today a number of proprietary chemicals known as NSP preservatives (NSP meaning "nonswelling and paintable"). These consist of a toxic chemical carried in nonaqueous volatile solvent with or without the further addition of water-repelling ingredients. Those without water-repellent consist typically of a solution of toxic in volatile petroleum solvent, such as mineral spirits or Stoddard's solvent, and in addition may contain a proportion of less volatile solvent such as fuel oil to facilitate spreading the preservative through the wooden structure.

After a ladder has satisfactorily passed the visual-acceptance examination, it should be heated in a kiln, oven, or warm room and then immersed for 3 min in the cool solution in a shallow pan, after which it is allowed to drain and dry. The solvent is

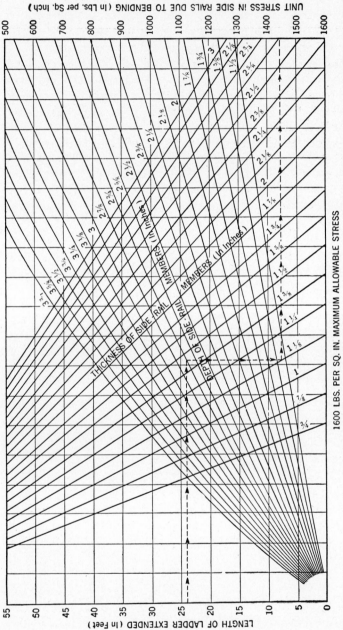

1600 LBS. PER SQ. IN. MAXIMUM ALLOWABLE STRESS

FIG. 1. Chart for determining the approximate stress in the side rails of wood ladders, based upon a 200-lb concentrated load applied at the center of the ladder. The stress should not exceed 1,600 lb per sq in. For ladders having recesses for treads or cleats, the **net** dimensions should be used.

flammable, but the treated ladder when dry is not any more combustible than before treating.

When thoroughly dry the ladder may, if so desired, be coated with paint, varnish, or enamel to improve its appearance, but these coatings will not noticeably increase the resistance of the wood to decay.

If a ladder has been thoroughly inspected prior to painting, the user can rest assured that no defects are likely to develop under a good coat of paint. For instance, knots, cross grain, and compression wood cannot develop after painting. Checks, if they develop, will surely show on the surface, as will subsequent compression failures. Decay and brashness due to exposure to heated air can be determined by jabbing with a knife, whether the ladder is painted or not.

Once a ladder has been approved by the inspector, it should be branded with the date as a mark of acceptance, and only such branded ladders should be used on the job. In fact, some companies brand serial numbers on all their ladders for identification and for record purposes. It is believed that, if only properly constructed ladders are allowed to be put in circulation, and if they are treated with reasonable care thereafter, the accidents due to ladder failure can be reduced to an insignificant figure.

The inspection of used wood ladders is largely a matter of common sense. For instance, anyone should know enough to condemn a ladder that has a broken side rail or a broken rung or tread. Nor should a ladder be used if it has a rung or other structural part missing, although we occasionally find such a ladder in use in spite of our rigid safety inspections. A ladder containing repairs made by any person other than the authorized ladder builder should not be used. We have all seen ladders with broken side rails reinforced by nailing a cleat across the break. Split rungs are frequently observed bound with wire.

Be on the lookout for loose rungs or treads that rotate more or less when grasped or stepped upon. Also watch for treads in stepladders that are slipping out of the gains or recesses in the side rails. Rungs that are worn down to $\frac{3}{4}$ in. minor diameter or treads such as on stepladders that are worn to less than $\frac{3}{8}$ in. at the nosing should be tagged and returned to the ladder shop for reconditioning. Notice if there are any badly worn spots such as are caused by the brackets upon which a ladder may have been hung, especially on a truck. In this case it will be a matter of using your best judgment to determine how much the wood may be worn away before it is unsafe for further use.

Such wear and also exposure to the weather will cause splinters, which may result in hand injuries. In some instances these splinters are soft and woolly and present only a slight hazard, but in most cases it is advisable to have the ladder sandpapered down.

In ladders used in street manholes, especially, we frequently find the ends of the side rails mushroomed. Where steel shrouds are not provided at the ends of the side rails, the wood sometimes brooms out so as to look like an old paintbrush. This is owing largely to the fact that the ladders stand in puddles of water in the manholes and absorb considerable moisture. Then, again, dropping the ladder carelessly into the manhole does not help matters. Mushrooming to a limited degree may act as a safety foot and help to prevent the ladder from slipping. However, when mushrooming develops, it indicates that decay has also set in, and we cannot tell how far up on the side rail it has progressed beneath the paint. Heavily painting or tarring the ends of the side rails, and care in lowering the ladders into manholes, will largely increase the life of the ladders.

We must watch for signs of mechanical injury such as bruises caused by dropping the ladder against some hard object or by the ladder's being run over or into by a vehicle, by being struck by a hammer, etc. Bruised spots are readily susceptible to dry rot, so when a bruise or other injury is observed, it is advisable to probe around this area with a sharp instrument. Soft, punky wood indicates rot, and the inspector will have to judge if the decay has advanced far enough to weaken the ladder, and if so, he should not hesitate to condemn it.

Of course, if any nails or screws project or are missing, they can be driven in or replaced, as the case may be, without returning the ladder to the shop. Occasionally we find a ladder with one side rail shorter than the other, thereby throwing it out of plumb. Owing to drying out of the moisture, especially in hot locations such as boiler rooms, the wood may shrink so that the rungs become loose in the side rails, causing the ladder to become rickety. A good way to determine the extent of such play is to stand the ladder on a level floor with the wide face of one side rail against a vertical wall or building column. Then while holding that side rail firmly against the wall, raise the other rail until the play is taken up.   If this distance is greater than $\frac{3}{8}$ in., the ladder certainly should be condemned.   In fact a much lower tolerance is desirable.

If unseasoned or improperly seasoned wood has been used in the construction of the ladder parts, the shrinkage may cause the ladder to warp sufficiently to make it unsafe for use.

If a test indicates very brash wood and the ladder is therefore condemned, it will be interesting to stand it up and then allow it to fall over and strike against some object near the floor level. The brash side rail may be expected to snap off in a clean splinterless fracture.

Single Ladders.   The discussion thus far has been general; that is, it applies to all types of ladders.   We shall now discuss the inspection of various types of ladders.   In addition to what has already been mentioned, there are a few more words to be said about ordinary straight ladders.   For instance, some straight ladders are provided with safety feet or safety hooks.   If the so-called "safety feet" are used, the abrasive or other antislip surface should be in good condition and not gummed up or coated with a foreign substance.   Most safety engineers have found that, although safety feet are in themselves a safety device, a great many men using ladders so equipped place undue confidence in them with the result that such ladders do slip.   For this reason the writer does not recommend the use of safety feet. Safety hooks, if provided, should be of the proper shape and length required to secure the ladder properly.

Extension Ladders.   Occasionally we find men using the sections of an extension ladder separately.   There is little objection to this provided the upper section of the ladder is used upside down so that the rung missing at the locks will be at the top of the ladder where it is less liable to cause an accident.

As far as its strength is concerned, it makes no difference which side of an extension ladder is placed toward the building.   But there is considerable disagreement on the question as to which way the ladder should be placed.   Perhaps 99 per cent of the extension ladders used on construction and painting jobs are placed with the upper, or "fly," section in front of the base section.   Most fire departments, including New York City, however, follow the practice recommended by the National Fire Protection Association, which specifies that the fly section be at the rear of the base section.

They point to several advantages of the NFPA method:

1. In descending the ladder, there is no missing rung near the bottom of the fly section to cause an accident.

2. When changing from the fly section to the base section, while descending the ladder, the rungs of the lower section appear to be farther out, and thus there is less danger of the foot's slipping off the rung.

3. When raising the fly section, the ladder is stood nearly vertical, but leaning slightly toward the building. Thus, the pull on the halyard largely balances the pull of gravity tending to upset the ladder.

On the other hand, the conventional use of the ladder has the halyard behind the ladder, where it is less likely to interfere with the feet when climbing.

Stepladders. Stepladders should bear evenly on all four legs on a plane surface or floor. They should have the truss rods under the treads or the step brackets in condition and drawn up tight. The diagonal bracing and the dowels on the rear section should be in good shape and not deformed by men using the rear of the ladder to climb up on. The spreader bars should operate properly and should not present unnecessary finger hazard. Rope spreaders should not be permitted.

The auxiliary step at the top of some platform ladders should be carefully examined, as the supporting members are usually very lightly constructed and are easily bent out of shape.

Sectional Ladders. Sectional ladders are not so common as the other types, but nevertheless many are in use, particularly by window cleaners. In addition to examining them as described in the general discussion above, special attention should be paid to the locking devices for holding the sections together. Whether these be of the friction-button, hook, or wing-nut type, they should operate properly and dependably. Look at the condition of the projecting ends of the top rung of the lower and intermediate sections that support the forked ends of the side rails of the sections above. Also note the condition of these forked ends of the side rails and their reinforcing plates.

Trestle Ladders. There is little to be said concerning trestle ladders in addition to what has already been discussed. The inspection should include the spreader bars, the locks for securing the extension section, and the wear on the rungs caused by these locks. Also, the ladder should bear on all four legs.

As mentioned previously, it is very difficult to draw an arbitrary line to differentiate between a safe and unsafe ladder. No two inspectors will agree as to just how far decay, for instance, may progress before the ladder is rendered unsafe for use. The value of a ladder inspection, therefore, depends entirely upon the judgment of the inspector with the help of a few rules such as just discussed.

A ladder of sufficient length shall always be selected for the work to be done. In general, the length of the ladder should be such that the work can be performed from at least the fourth rung from the top of the ladder. This will permit the side rails to be grasped conveniently. If the ladder is too short for the work at hand, get a longer ladder.

In selecting an extension ladder for a particular job, it should be remembered that this type of ladder is designated by its nominal length, which is the sum of the lengths of the sections. The usable length of the ladder is 3 to 10 ft less than the nominal length due to the overlap of the sections. This overlap is 3 ft on ladders up to and including 36 ft, 4 ft on 40- and 44-ft ladders, and 5 ft on longer ones.

Figure 2 will facilitate determining the nominal length of an extension ladder required to reach to a given height when placed at the proper angle.

Before using any ladder, thoroughly inspect it and be satisfied that it is in good condition.

Short ladders shall not be spliced together to make a long ladder.

Ladders shall not be used as guys, braces, and skids and (except as noted below) for scaffolding or for any purpose other than that for which they were intended.

An extension ladder can be used in a horizontal position as an improvised scaffold staging, provided

1. The ladder is in its collapsed or unextended condition.

2. The ends rest upon fixed supports, never upon suspended ropes. The supports shall be at least 12 in. from the ends of the ladder.

3. A thin plank or board is lashed to the rungs of the top section to form a flooring.

4. The load shall not exceed 300 lb concentrated at the center of the span or 600 lb uniformly distributed over its length.

When not in use, ladders should be stored in dry locations where they will not be exposed to the elements but where there is ample ventilation. They shall not be stored near radiators, steam pipes, stoves, or other places subjected to excessive heat.

Ladders stored in a horizontal position should be supported at a sufficient number of points to avoid sagging and permanent set.

Ladders carried on vehicles should be adequately supported on brackets so designed or padded as to avoid chafing of the side rails due to road shock.

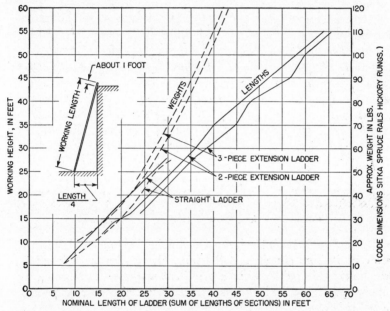

Fig. 2. Chart showing the nominal length and approximate weight of a ladder to reach to a given height.

Ladders that are damaged beyond repair should be destroyed or otherwise rendered useless. Those in need of minor repairs shall be sent to the carpenter shop for reconditioning.

When using a ladder, always place it at the proper angle, which is indicated when the horizontal distance from the base of the ladder to a point directly below the upper support is about one-fourth of the inclined length of the ladder from the base to the point of support. This is illustrated in Fig. 3. Rungs are always 12 in. apart, and this makes it easy to measure the length of the ladder.

It is poor practice to have the upper end of a ladder extend more than 3 or 4 ft above its upper support or to make it possible for a man to step onto a rung above the upper support, as this may cause the base of the ladder to "kick out."

When using an extension ladder, make certain that both locks are in good operating condition and engaged onto a rung of the lower section. Use the surplus length of the hoisting rope to lash together the adjacent rungs of the two sections. This will be an added safeguard in event of failure of the locks.

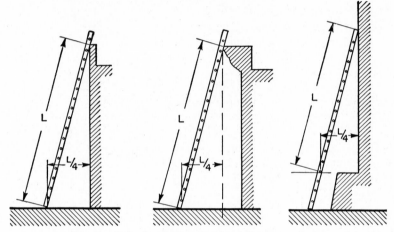

Fig. 3.   Rules for placing a ladder at the proper angle.

If practicable, lash the bottom of the ladder to a fixed object to prevent it from slipping. Otherwise, always have a helper hold the base of the ladder while a man is at work on it. Lashing the upper end of the ladder to the structure will also prevent possible upset sideways.

When a long ladder is to be raised, place the base against a wall, curbstone, or other fixed object or have a helper squat with his feet against the "heels" of the ladder side rails or on the bottom rung and grasp the second rung with his hands, meanwhile throwing his weight backward as far as possible. Then lift the upper end of the ladder over your head and walk toward the base of the ladder, grasping rung after rung until the ladder is in the vertical position. Then "walk" the ladder to the working position and carefully lower it against the wall.

To raise or lower ladders safely the following man power is recommended. Of course, by butting the base against a fixed object the "heel" men may be dispensed with.

Up to 18-ft straight ladder, 1 man to raise.

Up to 25-ft straight ladder, 1 man to raise and 1 man to heel.

Up to 30-ft straight ladder, 2 men to raise and 1 man to heel.

Up to 24-ft extension ladder, 1 man to raise.

Up to 30-ft extension ladder, 1 man to raise and 1 man to heel.

Up to 40-ft extension ladder, 2 men to raise and 1 man to heel.

Up to 55-ft extension ladder, 3 men to raise and 2 men to heel.

No more than two men shall be permitted on a straight, extension or stepladder at one time, and they shall not ascend or descend simultaneously.

Wood ladders should not under any condition be given a load test in excess of the normal load expected to be placed on same, as

TABLE II. COMPARATIVE WEIGHTS OF WOOD, ALUMINUM, AND MAGNESIUM LADDERS

Type	Length, ft	Wood, lb	Aluminum, lb	Magnesium, lb
Straight......	10	20	$18\frac{1}{2}$	9
	12	24	22	14
	14	28	$25\frac{1}{2}$	16
	16	32	$29\frac{1}{2}$	18
Extension	24	43	50	32
	32	58	$64\frac{1}{2}$	49
	36	72	72	55
	40	100	79	61

Note: Wood ladders: Chesebro-Whitman, "Gold Medal."
  Aluminum ladders: Mason Products Co., "Maproco."
  Magnesium ladders: White Aircraft Corp., "White-light."

there is far greater probability of weakening a good ladder than of detecting a weak ladder.

This chapter would not be complete without reference to aluminum and magnesium ladders. These straight ladders and extension ladders are stronger than wood ladders, are more readily inspected, will not be damaged by the elements nor by storage in heated locations, and may weigh less than wood ladders. Metal ladders are not standardized in design, as are wood ladders. Each manufacturer has his own type of construction, so each ladder must be judged on its own merits. If abused by careless handling, metal ladders usually give visual evidence of weakness more readily than do wood ladders in which treacherous compression failures (Chapter VI) may exist unobserved.

Some manufacturers are producing ladders made of materials

such as plastic with a balsa-wood core to make a product of increased strength and durability. However, the weight is reduced very little, if at all, while the cost is high. If the increased life is considered, the greater cost may be justified. At the present time the ideal ladder which incorporates high strength, limited deflection, light weight, long useful life, and low cost has yet to be produced.

Weights furnished by well-known manufacturers are as given in Table II.

Warning. Metal ladders should not be used near exposed electrical equipment, including overhead distribution wires, craneway trolley conductors, switchboards, etc.

CHAPTER XIV

STRENGTH CALCULATIONS FOR TIMBERS

The charts given in this chapter will prove helpful to the rigger in selecting the proper timbers to be used as beams and posts for supporting heavy loads. It should be understood that the strength of any species of wood varies considerably with the width of the annual rings, nature of defects, etc., so the loads given in these charts may not agree exactly with the results obtained from other sources.

There are a few general rules to remember concerning the safe loads and deflection of wood beams. For instance,

1. The safe load on a beam varies directly as the width. Thus, doubling the width of the beam doubles the safe load.

2. The safe load on a beam varies as the square of the depth. Thus, doubling the depth gives a safe load four times as great.

3. Placing two planks on top of each other gives twice the strength of one plank, but securely nailing or doweling their ends together to prevent slipping of one plank on the other gives four times the strength of one plank or twice the strength of the two planks placed on each other.

4. The safe load on a beam varies indirectly as the span. Hence, doubling the span cuts the safe load by one-half.

5. A beam of a given size, material, and span will carry a uniformly distributed load twice as great as a concentrated load applied at the center of the span (provided the unbraced span is relatively short).

6. The deflection of a beam of given size, material, and span varies directly as the load. Doubling the load doubles the deflection.

7. The deflection of a beam of given size, material, and loading varies as the cube of the span. Thus, doubling the span gives $2 \times 2 \times 2$ or eight times the deflection.

In order to show the use of the charts, it is desirable to work out some practical problems step by step, keeping in mind that the span is the distance from *center* to *center* of supports.

Problem 1. Given an 8 × 12 in. southern yellow pine beam of select grade, placed so as to have a span of 18 ft. Required to find the safe concentrated load.

Enter the chart (Fig. 1) at the bottom of "18-ft span," and

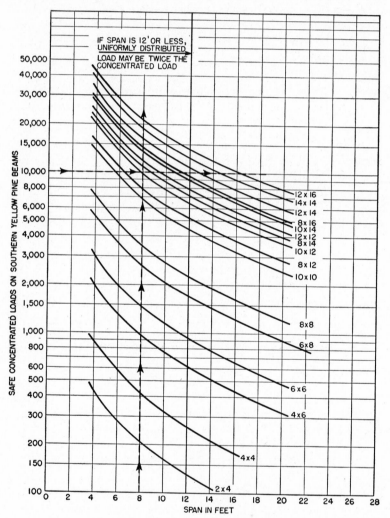

FIG. 1. Safe concentrated loads on dry wood beams of select-grade southern yellow pine. Larger dimension vertical. For other species use the following factors:

Douglas fir	1.00	Redwood	0.75
Oak	0.87	Spruce, eastern hemlock	0.70
West-coast hemlock	0.80	Southern yellow pine	1.00

(Note: Arrows on chart refer to Problem 3.)

follow upward to the curve representing an 8 × 12 beam, then follow to the left and read 3,250 lb safe concentrated load on the southern yellow pine beam. However, if the beam were of redwood we should multiply the 3,250 lb by 0.75 and obtain a safe load of 2,435 lb. (The strength of other species of wood relative to southern pine is given with the chart, Fig. 1.)

Problem 2. Given a 4 × 6 in. spruce needle beam on a 9-ft span supporting half of a scaffold. What uniform load will the scaffold safely carry?

Again, enter the bottom of the chart at 9-ft span, follow upward to the 4 × 6 curve, and read 850-lb safe concentrated load on a

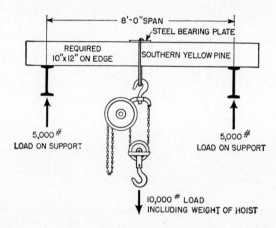

FIG. 2. Diagram of load on beam considered in Problem 3.

southern pine beam. Taking .70 times 850 gives 595-lb concentrated load on a spruce beam. The span being less than 12 ft, the safe uniformly distributed load is twice 595, or 1,190 lb, and inasmuch as two needle beams support the scaffold the safe load on the scaffold is twice 1,190, or 2,380 lb.

Problem 3. Given a concentrated load of 10,000 lb (including the weight of the chain hoist, Fig. 2) on a beam on an 8-ft span. What size southern pine beam is required?

Enter the chart from the bottom at 8-ft span, and draw a vertical line. Also, enter from the left at 10,000 lb, and draw a horizontal line (see lines drawn on Fig. 1) until it crosses the vertical line, which is almost on the 10×12 curve. Thus, we can use the 10×12 beam or any beam of larger size (above it on the chart).

For any other species, first divide the load by the factor given below Fig. 1 and then enter the chart from the left by using this corrected figure.

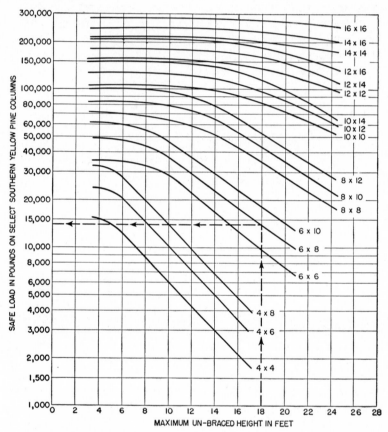

FIG. 3.   Safe loads on dry wood columns or posts of select grade lumber.   Values indicated are for select southern yellow pine.   For other species use the following factors:

Douglas fir (coast region) 1.00	Spruce.............. 0.62
Oak and redwood...... 0.78	Eastern hemlock....... 0.55
West coast hemlock.... 0.70	Southern yellow pine... 1.00

Likewise, we can work out some problems in the use of wood posts or columns.   In such cases the unbraced height of the column divided by the least dimension of the timber governs to a large degree the load the column may safely carry.

Problem 4. Given a 6 × 8 in. southern yellow pine post having an unbraced height of 18 ft. What load will it safely support?

Enter the chart (Fig. 3) at the bottom at 18 ft, follow up to the 6 × 8 curve, and then follow to the left and read 14,000-lb safe load. If any other species of wood were used, the 14,000-lb load would then be multiplied by the factors below the chart.

Problem 5. Given a 4 × 6 Douglas fir post 10 ft high. What load will it safely support? Also, what is the safe load if the additional bracing shown in Fig. 4 is added to reduce the unbraced height to 5 ft?

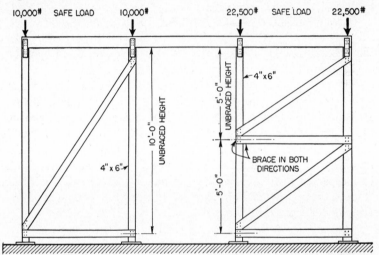

Fɪɢ. 4. Loading of poorly braced and well-braced wood posts considered in Problem 5.

Enter the chart at the bottom at 10 ft, follow upward to 4 × 6, then left, and read 10,000-lb safe load for select southern pine. Inasmuch as Douglas fir has the same strength, the safe load is still 10,000 lb.

Again let us enter the chart at 5 ft, follow upward to 4 × 6, and we read 22,500-lb safe load for southern pine (and for Douglas fir). This should emphasize the importance of adequate bracing on all posts and columns.

Problem 6. Given a load of 100,000 lb to be carried on a west coast hemlock post 20 ft high between braces. What size timber should be used?

From the table below Fig. 3, we find that west coast hemlock has a strength of 0.70 times that for southern pine. In other words if a west coast hemlock timber would support 100,000 lb, then a southern pine of the same size would carry $100,000 \div 0.70 = 142,800$ lb. Hence, we enter the chart at the left at 142,800 lb and at the bottom at 20 ft. These lines intersect slightly below the 12×14 curve, so a 12×14 or any larger size will suffice.

However, if we want to economize on the size of timber used, let us assume that the posts are braced 10 ft above the ground. Thus, entering the chart at 142,800 lb and at 10 ft, we find that a 10×14 post will safely carry the load. The difference is more noticeable on the smaller sizes with relatively longer unsupported lengths.

Of course all posts should have square ends and should rest on sills to distribute the load over a larger area of the ground. Also, the bracing should be adequately nailed or bolted to the posts in order to provide the required rigidity.

The mathematical calculation of timber beams is quite involved, owing not only to the effect of the direct load on the beam but also to the presence or absence of horizontal bracing to prevent the beam from deflecting sideways and then rolling over on its side and failing. And, in addition, on short deep beams there is a tendency for the beam to split along its neutral plane and fail like two separate beams or planks placed one on top of the other. Hence, the mathematical calculations are very complicated and are omitted from this chapter.

CHAPTER XV

STRENGTH CALCULATIONS FOR METAL BEAMS

Before proceeding with the discussion of steel and light metal beams, it might be well to describe briefly the standard structural shapes. In standard WF (meaning "wide flange") beams, I beams, and channels the vertical central portion is called the "web" while the horizontal top and bottom parts are called the "flanges." In the WF beams the flanges are, as the name indicates, wide in relation to the depth of the beam. Also these flanges are of uniform thickness.

American standard I beams have relatively narrow flanges which are tapered, being thinner at the "toe" of the flange. Channels appear like half of an I beam, the flanges being on only one side of the web.

In locating WF or I beams, the dimension indicates the center of the web. Channels are located by measuring to the back of the web or "heel" of the flanges. The span of a simple beam supported at or near the ends is the distance between supports and is always *measured in inches*. The length of the span depends to a degree upon the nature of the material upon which the beam rests. Figure 1 shows simple beams with their deflections exaggerated for clarity. If the supports are very hard, such as heavy steel or concrete, the effective span is from edge to edge of supports (Fig. 1a). On the other hand, if the beam rests upon timbers that compress slightly under the bearing load (Fig. 1b), or if the steel beams upon which it rests can rotate slightly under the beam (Fig. 1c), then the span is measured from the center of one bearing to the center of the other bearing. When selecting a metal beam, keep in mind the following rules:

1. A beam of given size and span will carry a uniformly distributed load twice as great as a load concentrated at the center of the span. In other words, if from the table we find that a certain beam on a given span will safely carry a concentrated load of 7,500 lb, then if the load is spread out over the entire length of the beam, 15,000 lb may be safely applied. Likewise, on a cantilever beam

the allowable distributed load is twice the allowable concentrated load.

2. The safe load on a simple beam varies indirectly as the span. Hence, with a beam of a given size if we double the span, we reduce the safe load to one-half. Likewise, on a cantilever beam if the lever arm is reduced to one-half the length, the load may be doubled.

3. The deflection or bending of a simple or a cantilever beam under load varies directly as the load. Thus, doubling the load will ordinarily double the deflection.

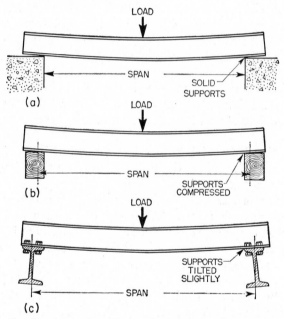

Fig. 1.   Sketches showing how the length of the span of a beam is measured.

4. The deflection of a simple beam of a given size and load varies as the cube of the span. Thus, if we multiply the span by 2, then we must multiply the deflection by 2^3, or $2 \times 2 \times 2 = 8$. If we multiply the span by $2\frac{1}{2}$, then the deflection is $(2\frac{1}{2})^3$, or $2\frac{1}{2} \times 2\frac{1}{2} \times 2\frac{1}{2}$ or $15\frac{5}{8}$ times as great.

5. The load applied to a beam must include not only the useful load to be lifted but also the weight of the slings, the hoist tackle or chain hoist, the pull on the hauling line or on the hand chain, and (if the beam is very heavy) the weight of the beam itself.

TABLE I. STRENGTH OF STEEL, ALUMINUM, AND MAGNESIUM BEAMS *

A. STEEL WF BEAMS 18,000 lb per sq in.			*A.* STEEL WF BEAMS 18,000 lb per sq in.		
Depth × width, in.†	Maximum bending moment	Weight per ft, lb	Depth × width, in.†	Maximum bending moment	Weight per ft, lb
36 × 16½	15,000,000	230	16 × 8½	1,692,000	58
33 × 15¾	12,070,000	200	14 × 10	1,660,000	61
30 × 15	9,500,000	172	18 × 7½	1,602,000	50
36 × 12	9,050,000	150	12 × 12	1,583,000	65
33 × 11½	7,300,000	130	12 × 10	1,273,000	53
27 × 14	7,260,000	145	14 × 8	1,139,000	43
24 × 14	5,950,000	130	16 × 7	1,014,000	36
30 × 10½	5,900,000	108	10 × 10	983,000	49
21 × 13	4,500,000	112	12 × 8	934,000	40
24 × 12	4,480,000	100	14 × 6¾	752,000	30
27 × 10	4,380,000	94	10 × 8	630,000	33
14 × 16‡	4,085,000	142	12 × 6½	614,000	27
18 × 11¾	3,310,000	96	8 × 8	493,000	31
21 × 9	3,250,000	82	10 × 5¾	387,000	21
24 × 9	3,150,000	76	8 × 6½	375,000	24
16 × 11½	2,720,000	88	8 × 5¼	254,000	17
14 × 14½	2,485,000	87	6 × 6	182,000	15½
21 × 8¼	2,270,000	62	5 × 5	153,000	16
14 × 12	2,180,000	78	4 × 4	75,000	10
18 × 8¾	2,110,000	64			

B. STEEL I BEAMS 18,000 lb per sq in.			*C.* MAGNESIUM I BEAMS Alloy J–1 or 0–1 10,000 lb per sq in.		
Depth × width, in.†	Maximum bending moment	Weight per ft, lb	Depth × width, in.†	Maximum bending moment	Weight per ft, lb
24 × 7⅞	4,210,000	106	6 × 3⅝	73,600	2.81
24 × 7	3,130,000	80	5 × 3	49,000	2.24
20 × 7	2,700,000	85	4 × 2⅝	30,300	1.73
20 × 6¼	2,180,000	65	3 × 2⅜	16,800	1.28
18 × 6	1,590,000	55			
15 × 5½	1,060,000	43			
12 × 5¼	806,000	41			
12 × 5	648,000	32			
10 × 4⅝	439,500	25	*D.* MAGNESIUM H BEAMS Alloy J–1 or 0–1 10,000 lb per sq in.		
8 × 4	255,500	18			
7 × 3⅝	187,000	15	Depth × width, in.†	Maximum bending moment	Weight per ft, lb
6 × 3⅜	131,500	12½			
5 × 3	86,300	10	6 × 6	146,900	5.10
4 × 2⅝	54,000	8	5 × 5	95,300	4.21
3 × 2⅜	30,600	6	4 × 4	53,600	3.07

* Tables are arranged in the order of their bending moments.
† "Nominal size ' is depth by width.
‡ Actual depth varies considerably.

E. ALUMINUM H BEAMS Alloy 17ST 15,000 lb per sq in.		
Nominal size, in.	Maximum bend- ing moment	Weight per ft, lb
8 × 8	425,000	11.51
6 × 6	220,000	8.04
5 × 5	143,000	6.63
4 × 4	80,500	4.85

F. STEEL JOISTS 18,000 lb per sq in.		
Nominal size, in.	Maximum bend- ing moment	Weight per ft, lb
12 × 4	226,500	14
10 × 4	189,000	$11\frac{1}{2}$
8 × 4	140,000	10
6 × 4	91,100	$8\frac{1}{2}$

G. STEEL CHANNEL BEAMS 18,000 lb per sq in.		
Nominal size, in.	Maximum bend- ing moment	Weight per ft, lb
18 × 4	1,100,000	43
15 × $3\frac{3}{8}$	750,000	34
12 × 3	385,000	21
10 × $2\frac{5}{8}$	241,500	15
9 × $2\frac{1}{2}$	189,200	13
8 × $2\frac{1}{4}$	145,700	$11\frac{1}{2}$
7 × $2\frac{1}{8}$	108,000	10
6 × 2	77,500	8
5 × $1\frac{3}{4}$	54,000	7
4 × $1\frac{5}{8}$	34,200	5
3 × $1\frac{1}{2}$	19,800	4

H. ALUMINUM I BEAMS Alloy 17ST 15,000 lb per sq in.		
Nominal size, in.	Maximum bend- ing moment	Weight per ft, lb
12 × 5	425,000	11.31
10 × $4\frac{5}{8}$	296,000	9.01
9 × $4\frac{5}{16}$	229,500	7.72
8 × 4	173,000	6.53
7 × $3\frac{5}{8}$	136,000	5.42
6 × $3\frac{3}{8}$	88,500	4.43
5 × 3	59,000	3.53
4 × $2\frac{5}{8}$	36,400	2.72
3 × $2\frac{3}{8}$	20,200	2.02

Calculations for steel or light metal beams are not too difficult if certain facts are known. For instance,

Known	*Required*
Span of beam, size of beam, material................Safe load	
Span of beam, load, material........................Size of beam	
Size of beam, load, material........................Maximum span	

Inasmuch as the same unit stress (18,000 lb per sq in.) is used for all steel beams, Table I lists the product of the unit stress and the section modulus under the heading of maximum bending moment. This saves one step in the calculations.

Figure 2 gives the formulas for determining the bending moment when the size of the beam is to be calculated. Conversely, if the beam size is known, the formulas may be used to find the maximum span or load. In all instances

M = bending moment, lb-in. (lb multiplied by in.)

P = concentrated load, lb

W = total uniformly distributed load, lb

L = length of span, in.

Perhaps the best way to explain the method of making the calculations is to work out some practical problems. In all instances unless the beam is very narrow relative to the span, the calculations are quite simple. But on beams with a high "slenderness ratio" additional investigation must be made. The slenderness ratio is the length of the span between lateral bracing of the compression flange divided by the width of the beam flanges.

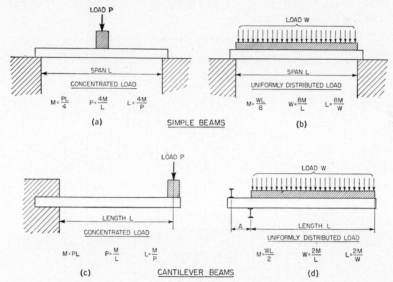

FIG. 2. Formulas for calculating the bending moment of a beam under various types of loading.

The reason we are interested in the slenderness ratio is that, if a beam is very slender and an excessive load is applied vertically, there is a tendency for the beam to deflect sideways, then twist and roll over on its side. The beam, being much weaker on its side, will fail and allow the load to drop. Bracing the compression flange against deflection sideways, so that the slenderness ratio of the longest unbraced part of the span is 15 or less, will prevent this from happening, and the full calculated load can be safely carried.

Problem 1. Given a steel beam that measures 12 in. deep with an 8 in. wide flange and used on a span of 9 ft (108 in.). What load concentrated at the center of the span will it safely support?

The dimensions indicate a 12 × 8 WF beam. (All beams are manufactured in several thicknesses and weights, but as the lightest weight beams of each size are most readily obtainable, only these

are listed in the table.)   From Table IA we find that the allowable
bending moment for a 12 × 8 WF steel beam is 934,000 lb-in.   In
Fig. 2a we find that

$$P = \frac{4M}{L} = \frac{4 \times 934,000}{108} = 34{,}590\text{-lb safe load}$$

But before we accept that figure as final let us check the slenderness ratio.   108/8 = 13.5.   Referring to Fig. 3, we find that the
full load can be carried.

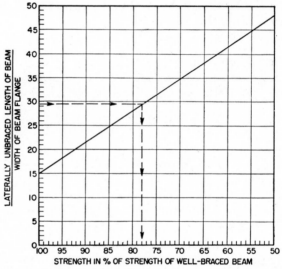

FIG. 3.   Chart for estimating the reduction in strength of a steel beam due to a
long span unbraced laterally.

Problem 2.   Given a load of 22,000 lb uniformly distributed
along the length of a simple beam having a span of 16 ft 0 in.
(192 in.).   What size beam is required?

Using the formula in Fig. 2b, we find that

$$M = \frac{WL}{8} = \frac{22{,}000 \times 192}{8} = 528{,}000\text{ lb-in.}$$

Looking in Table I, any beam having an allowable bending moment
of 528,000 lb-in. or more will be acceptable.   We shall select the
12 × 6$\frac{1}{2}$ in. beam (indicated in Table IA), which has an allowable
bending moment of 614,000 lb-in.

Now let us investigate the slenderness ratio.   192/6$\frac{1}{2}$ = 29$\frac{1}{2}$.
Referring to Fig. 3, we enter the chart from the left at 29$\frac{1}{2}$, then

read at the bottom 78 per cent. So taking 78 per cent of 614,000 lb-in. we get 478,000 lb-in. This is less than the 529,100 lb-in. required, so this beam is not strong enough.

The next heavier beam will be investigated. The 10 × 8 in. WF beam is good for a bending moment of 630,000 lb-in. The slenderness ratio is 192/8 = 24. From the chart we find that the safe load is 86 per cent of 630,000 lb-in., or 542,000 lb-in. Thus, the 10 × 8 WF beam is acceptable.

Problem 3. Given a concentrated load of 12,000 lb and a 10 × 5¾ WF beam. What is the longest span that can be had and yet safely support this load?

From Table IA we find that the 10×5¾ beam is good for a bending moment of 387,000 lb-in. From Fig. 2a we find that $L = \dfrac{4M}{P} = \dfrac{4 \times 387,000}{12,000} = 129$ in. or 10 ft 9 in., provided, of course, that the slenderness ratio permits. The slenderness ratio is $\dfrac{129 \text{ in.}}{5\frac{3}{4} \text{ in.}} = 22\frac{1}{2}$, and the allowable load (from Fig. 3) is 88½ per cent of 12,000 lb, or 10,620 lb. The load is definitely 12,000 lb, so the span must be shortened and the calculations repeated.

An alternative, providing, of course, that the maximum possible span is desired, is to install rigid lateral braces against the top flange of the beam at the middle of the span to prevent it from deflecting sideways. If this is done, the slenderness ratio will be $\dfrac{64\frac{1}{2}}{5\frac{3}{4}} = 11.2$ and the full 12,000-lb load can be safely carried.

Problem 4. Given an outrigger scaffold 3 ft 0 in. wide, supported by 7-in. aluminum I beam outriggers spaced 8 ft 0 in. apart. The scaffold deck is made up of four 2 × 9 in. spruce planks laid abreast of each other. The general design of the scaffold is as shown in Fig. 4. What safe load can it carry, in pounds per square foot of scaffold deck?

The strength of this scaffold may be limited by
1. The strength of the deck planks.
2. The strength of the outriggers.
3. The strength of the anchorages for the outriggers.

First, let us consider the strength of the planks. From Fig. 2 in Chap. VII we find that a 2 × 9 spruce plank on an 8-ft span is good for about 116-lb load concentrated at the center of the span or double that amount (232 lb) if the load is uniformly distributed over the length of the span. The safe load being 232 lb for a plank 9 in. ($\frac{3}{4}$ ft) wide by 8-ft span, then $\dfrac{232 \text{ lb}}{8 \times \frac{3}{4}} = 38\frac{3}{4}$ lb per sq ft of floor area.

Next let us check on the strength of the outriggers. From Table I*H* we find that the allowable bending moment for a 7-in.

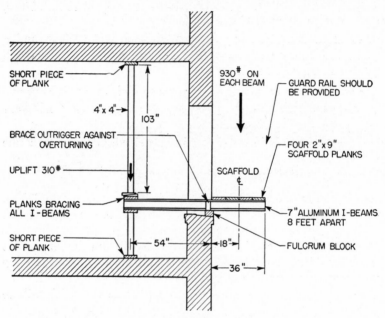

SHORT PIECE OF PLANK

930[#] ON EACH BEAM

GUARD RAIL SHOULD BE PROVIDED

4" x 4"→

103"

BRACE OUTRIGGER AGAINST OVERTURNING

FOUR 2"x9" SCAFFOLD PLANKS

UPLIFT 310[#]

SCAFFOLD ₡

PLANKS BRACING ALL I-BEAMS

7" ALUMINUM I-BEAMS 8 FEET APART

SHORT PIECE OF PLANK

54"

18"→

FULCRUM BLOCK

36"→

FIG. 4. Sketch of the scaffold outrigger considered in Problem 4.

aluminum I beam is 136,000 lb-in. And from Fig. 2*d* we note that $W = \dfrac{2M}{L} = \dfrac{2 \times 136,000}{36} = 7{,}550$-lb safe uniform load on each outrigger. As each beam supports a portion of the scaffold 3 ft wide by 8 ft long (4 ft on either side of the outrigger), or 24 sq ft, then $\dfrac{7{,}550 \text{ lb}}{24} = 315$ lb per sq ft of floor area.

The outriggers are therefore eight times as strong as the planks. Hence, the strength of the planking governs, and the scaffold is good for a uniform load of $38\frac{3}{4}$ lb on each square foot of floor area. From this should be deducted about 5 lb per sq ft for the weight of the planks themselves. Or let us say that the safe live load is 34 lb per sq ft.

But before finally agreeing on that figure we should check the anchorages for the outriggers. Having a gross load of $38\frac{3}{4}$ lb per sq ft and 24 sq ft of deck area, the actual load on each outrigger is 930 lb distributed uniformly. Or this may be considered concentrated at the middle of the cantilever or 18 in. from the fulcrum on the window sill. Referring to Fig. 6*b* in Chap. I, we find that 930 lb $\times$ 18 in. equals the uplift times 54 in. So $\dfrac{930 \times 18}{54} = $ 310-lb uplift, which is resisted by a 4 $\times$ 4 spruce post 103 in. high.

Referring to Fig. 3 in Chap. XIV we find that the safe load is 7,700 lb on the post whereas the actual load is only 310 lb. In other words, the post is of more than adequate strength to support the maximum load that can be supported by the scaffold planks, and the safe load is about 34 lb per sq ft as mentioned above.

For those who may wish to pursue this subject further the *Handbook* of the American Institute of Steel Construction is a good, and inexpensive, investment.

CHAPTER XVI

CRANES, HOISTS, AND DERRICKS

There is almost no limit to the number of different designs of cranes that can be built. Basically, a crane is described as a piece of hoisting equipment designed to pick up a load, transport it a reasonable distance, and land it again. A so-called "hoist" may be a complete hoisting unit mounted overhead in a fixed location or suspended from a small trolley attached to an I-beam truck. Or the term "hoist" can be applied to the power-driven mechanism and drums that are used in conjunction with a derrick, gin pole, or even a material-hoisting elevator.

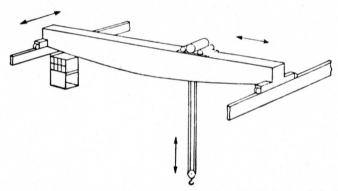

Fig. 1. Overhead or traveling crane. (*American Standards Association.*)

Traveling Types of Cranes. The traveling crane, gantry, half-gantry, bridge, and wall cranes operate on fixed straight tracks and usually have their hoist mechanism mounted on a movable trolley. These types are shown diagrammatically on Figs. 1 to 5 inclusive.* The hammerhead, pillar, and jib cranes have a fixed location with a boom that rotates or slews around it (Figs. 6 to 10). The portal crane is a pillar crane mounted on a gantry (Fig. 11).

These cranes and the monorail hoist as well are all equipped with two brakes on the hoist motion, usually a solenoid-operated

* Figures 1 to 18 inclusive are reproduced through the courtesy of the American Standards Association.

brake on the motor shaft and a mechanical load brake on one of the intermediate gear shafts.   Each brake should hold at rest a load of $1\frac{1}{2}$ times the rated load.   These cranes always have a limit device on the hoist motion to stop the motor automatically in the event the operator fails to shut off the power when the lower load block reaches the safe limit of travel.   The limit device is installed only as a safety feature, and it should not be depended upon in place of alertness on the part of the operator to shut off the power during normal operations.

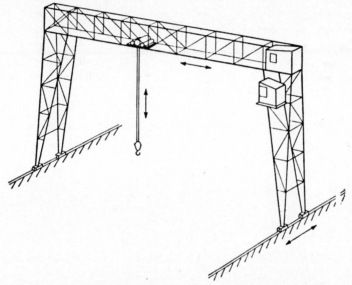

FIG. 2.   Gantry or bridge crane.   (*American Standards Association.*)

Before starting operations at the beginning of the day's work always pick up a capacity load a foot off the ground to test the brakes.   Of course, on very large cranes this is impracticable, for capacity loads are not always available.   Lower the load an inch or less at a time, and observe the drift, if any, due to faulty brakes.

The operator should always accelerate and decelerate the bridge and trolley motions slowly in order to avoid unnecessarily swinging the load.   Where a crane handles a large number of small loads daily, such as in a storehouse, the operator may learn by experience how to stop his crane quite quickly by aiming at a spot perhaps a foot short of the location where the load is to be placed; then as the load swings ahead of the crane, he releases the foot brake and

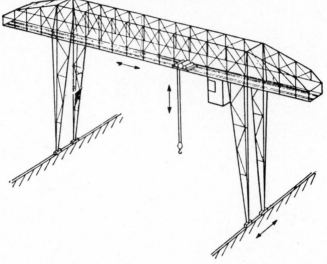

Fig. 3.   Cantilever gantry or bridge crane.   (*American Standards Association.*)

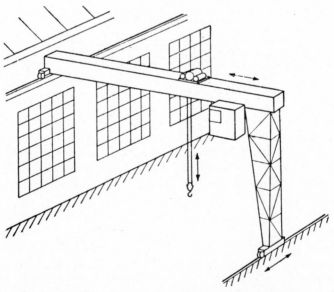

Fig. 4.   Half-gantry crane.   (*American Standards Association.*)

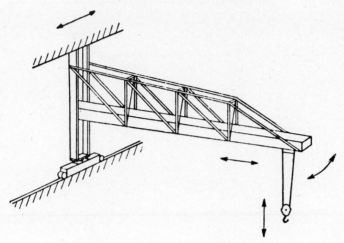

Fig. 5. Wall crane. (*American Standards Association.*)

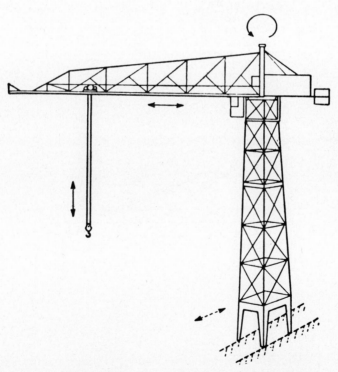

Fig. 6. Hammer-head crane. (*American Standards Association.*)

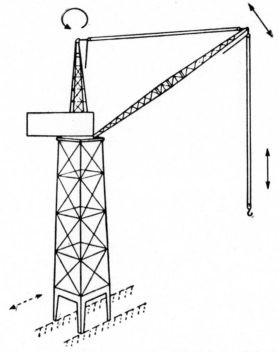

Fig. 7.   Tower crane.   (*American Standards Association.*)

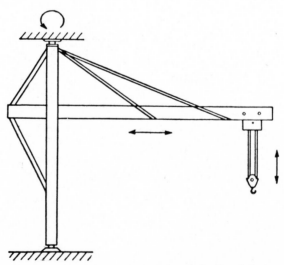

Fig. 8.   Jib crane.   (*American Standards Association.*)

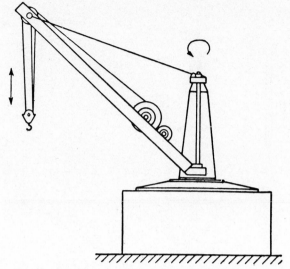

FIG. 9.   Pillar crane.   (*American Standards Association.*)

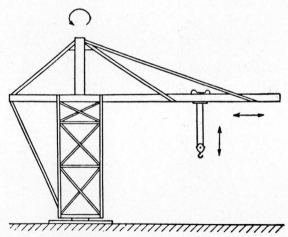

FIG. 10.   Pillar jib crane.   (*American Standards Association.*)

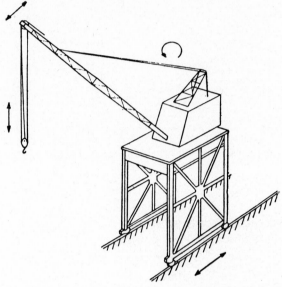

FIG. 11. Portal crane. (*American Standards Association.*)

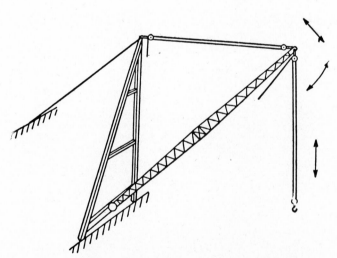

FIG. 12. A-frame derrick. (*American Standards Association.*)

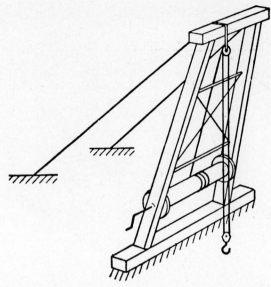

FIG. 13.   Breast derrick.   (*American Standards Association.*)

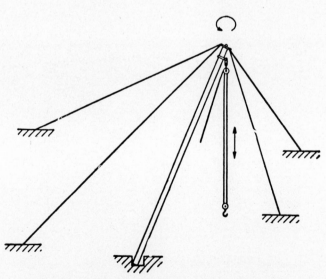

FIG. 14.   Gin pole.   (*American Standards Association.*)

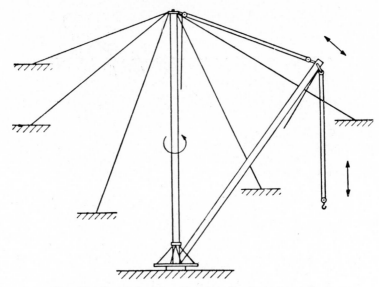

Fig. 15. Guyed derrick. (*American Standards Association*.)

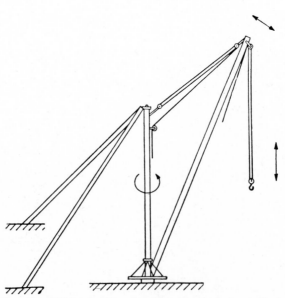

Fig. 16. Stiff-leg derrick. (*American Standards Association*.)

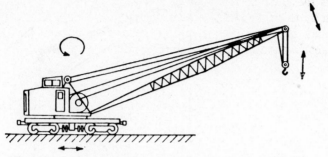

FIG. 17.   Locomotive crane.   (*American Standards Association.*)

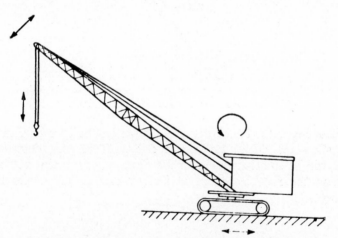

FIG. 18.   Crawler crane.   (*American Standards Association.*)

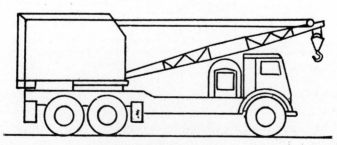

FIG. 19.   Automotive crane.

allows the crane to move ahead and come to rest over the load, which by this time has stopped swinging.

Cranes of these types should be thoroughly inspected at least once each month. Before starting the inspection, have the

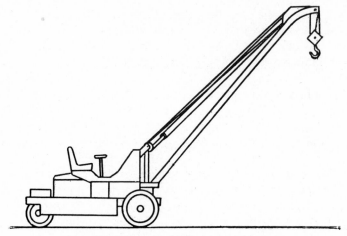

FIG. 20.   Three-wheel industrial truck crane.

operator lower the load block and land it on the floor or ground and allow the cables to become slack. If properly installed, there will be sufficient cable on the drum to permit this without bending the rope backward at its anchorage. Then open the

FIG. 21.   Fork-lift industrial truck.

power switch, and apply your own personal padlock to prevent anyone from accidentally energizing the crane while the inspection is in progress. If a lock cannot be applied, attach a warning sign to the main switch.

First examine the structure of the crane for bent or damaged members, corrosion, loose rivets and bolts, etc. Look at the bridge end-truck wheels for flat spots or worn or broken flanges. Check the alignment at the runway rails, and observe if the crane bridge is square with the runway.

Inspect every foot of hoist cable, giving special attention to that part which bears on the equalizing sheave, as this is often the first part of the rope to show evidence of failure. Sliding a rag along the cable will frequently help find broken wires, as it will usually catch on them. Note the condition of the cable anchorages on the drum, and look for broken wires where the rope is bent sharply to enter the hole. Also observe if the sheave grooves are of proper size for the rope and if the sheave and drum grooves have been corrugated so as to wear the rope excessively. Make sure that the cables are properly lubricated.

Try operating the limit device manually to make sure that it works freely. Pry open the solenoid brake to observe if the lining is coated with oil or grease. Open up the gear case, and ascertain that there is sufficient oil in it, and at the same time observe the condition of the mechanical load brake. The load brake is a special brake that prevents a descending load from turning over the hoist mechanism and motor. It is frequently of the disk type with an automatic retaining band G, which grips the brake wheel A (Fig. 22) and prevents it from turning in the lowering direction. Asbestos friction material is used for lining the disks as well as the retaining band. In operation the load tends to revolve the load brake pinion F. This pinion, in turning against the helix D, is thrust endwise on the shaft and forces the disk B against the brake wheel A, which is automatically held from revolving by the retaining band G. In this manner the hoisted load is locked and held from lowering.

To lower the load, the motor is run in the lowering direction whereby the helix D is released and the pressure of the disks B and C on brake wheel A is decreased sufficiently to allow slipping between these parts. This permits the load to be lowered. Increased motor torque and speed tend to open the brake against the load reaction, increasing the lowering speed, while a decrease in motor torque and speed is overcome by the load reaction, and the movement in lowering slows down or stops.

Observe the condition of the electric current collectors, trolley wires or collector rails, controllers, switchboard, and motors.

Then, if everything so far has been found to be satisfactory, make sure that all persons are clear, close the main power switch, and proceed with the test of the crane.   Try the foot-operated bridge brake, and make sure that the pedal does not strike the floor.   Then travel the bridge at about half speed, and apply the brake.   Repeat

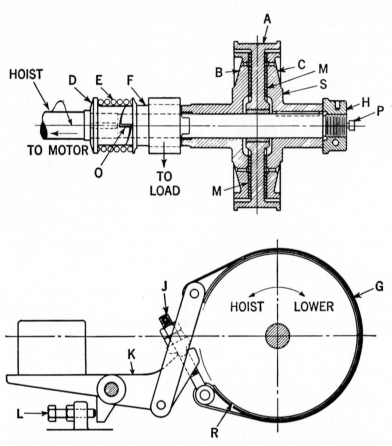

FIG. 22.   Mechanical load brake of typical design.   (*Harnischfeger Corp.*)

in the opposite direction.   Run the crane the entire length of the runway, and note any irregularities in the track, any unusual vibration, or binding of the wheels on the rails.   At each end of the runway check if the bridge is square with the runway by gently striking the rail stops.

Next test the trolley, remembering that very few cranes have brakes on this motion. Move it part way across the bridge, and stop it as gently as possible by "plugging" or reversing the motor.

Finally, test the hoist mechanism by running the load block up and down several times and observing the "feel" and the sound of the mechanism. Then carefully run the block up until it causes the limit device to trip out. Repeat at a somewhat higher speed and finally at full speed to make sure that the drift will not permit the block to strike the trolley structure. Make certain that the limit device is so wired that it will *open* the circuit rather than close it when functioning.

Next, have the regular operator take over the controls while you proceed to a position on the trolley. As the operator raises and lowers the load block at your direction, observe the action of the solenoid brake. (Make certain of your footing on the trolley and that you are clear of all moving parts. Also, keep your head low so as not to strike an overhead beam or roof truss should the bridge be moved unexpectedly.)

Block the solenoid brake in the "off" position, and have the operator lift a near-capacity load. The mechanical load brake alone must hold the load. Then, if the construction of the particular crane will permit, block off the load brake by disengaging the pawl or the band brake, and lift the load again. The solenoid brake alone must hold the capacity load.

Check the warning gong or horn, the fire extinguisher, condition of the rubber floor mat, etc. If it is an outdoor crane, make sure that the rail clamps or other anchorages against the wind are in good condition.

If the crane is floor-operated, it can be controlled by a pendent push-button control box or by means of six drop cords with handles on them. In order to prevent accidentally grasping the wrong handle when the operator is also acting as the rigger, it is suggested that each handle be different from the others as a means of identification. Figure 23 shows recommended shapes for the handles.

Mobile Cranes. The movable or mobile cranes include the locomotive crane, crawler crane, automobile-truck crane, and the smaller gasoline or storage battery cranes (Fig. 24). To inspect these cranes, have the operator lower the boom nearly horizontally or until the load block rests upon the ground; then stop the engine. Examine the boom carefully, both from the ground and by walking out on it. Strike all rivet heads with a machinist's hammer to detect any loose ones. Note all bent structural mem-

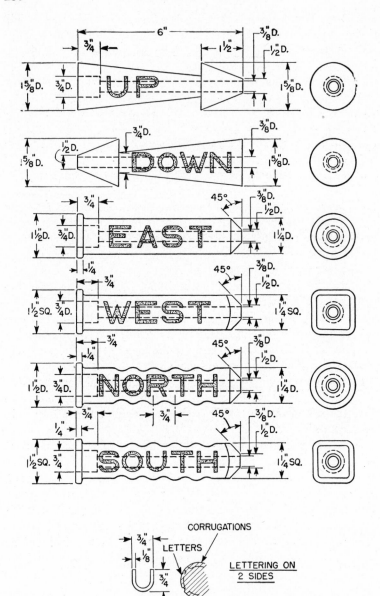

FIG. 23. Suggested shape of handles for the floor-operated cranes and hoists
With each handle distinguished by its shape, it is not necessary to see the handle
to identify it.

bers (legs, lattice members, gusset plates, etc.) and any parts worn by the cables. Check for excessive corrosion. Also inspect the structural members that form the anchorage for the boom hoist cable. Note any loose bolts.

Strike each sheave with the hammer to detect any cracks. Of course, check the cables, anchorages, etc. Inspect the crane engine and hoist mechanism, paying attention to loose or worn pins, keys, cotter pins, broken gear teeth, etc. Check the running gear, including wheels, crawler treads, axles, gears, sprockets,

FIG. 24.   Rotary type crane used in storage yard.

turntable, rollers, center pin, and other vital parts. Make sure that the rail clamps and/or out-riggers are in good condition.

Have the operator start up the engine and raise the boom to the normal operating position, then pick up a fair-size load and test the brake and frictions. Check the latch (if provided) on the foot brakes to ensure positive holding.

Observe the capacity plate or chart on the crane, making note of the safe lifting capacity at various radii. Keep in mind that the radius is measured from the center pin, *not* from the hinge pin. If weights of known magnitude are available, pick them up at the maximum safe radius, handle them at normal operating speeds,

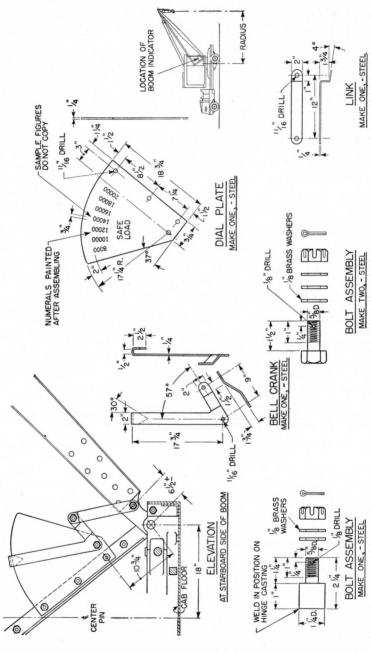

FIG. 25. Shop details for boom-indicating device to show the safe load with the boom in any position. (*Consolidated Edison Co. of New York, Inc.*)

NOTES: (1) Verify all dimensions of the crane before fabricating the indicator. (2) After assembling the indicator, ascertain from the manufacturer of the crane the radii at which each additional 1,000 lb may safely be lifted. Raise or lower the boom to give the desired radius and paint figures on the dial at the proper point. (3) Note that the boom radius is from the center pin of the crane, not front hinge pin.

282

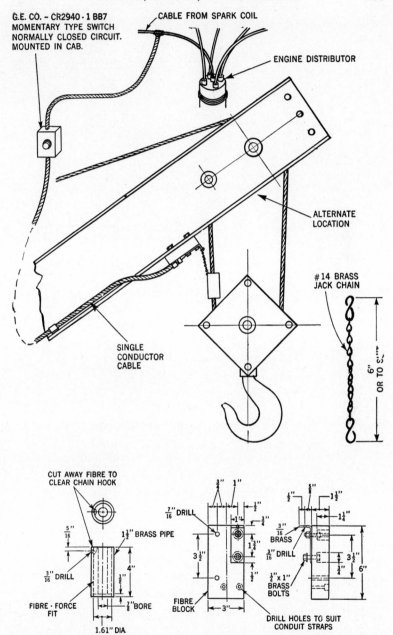

G.E. CO. – CR2940 · 1 BB7
MOMENTARY TYPE SWITCH
NORMALLY CLOSED CIRCUIT.
MOUNTED IN CAB.

CABLE FROM SPARK COIL

ENGINE DISTRIBUTOR

ALTERNATE
LOCATION

#14 BRASS
JACK CHAIN

SINGLE
CONDUCTOR
CABLE

6" OR TO SUIT

CUT AWAY FIBRE TO
CLEAR CHAIN HOOK

$\frac{5}{16}$"

$1\frac{1}{2}$" BRASS PIPE

$\frac{3}{16}$" DRILL

4"

$1\frac{1}{2}$"

FIBRE · FORCE
FIT

$\frac{7}{8}$" BORE

1.61" DIA.

$\frac{7}{16}$" DRILL

$\frac{3}{4}$"   1"

$\frac{1}{2}$"

$1\frac{1}{4}$"

$\frac{3}{4}$"

$1\frac{3}{4}$"

$3\frac{1}{2}$"

$\frac{1}{2}$"

FIBRE
BLOCK

3"

$\frac{1}{2}$" $\frac{5}{8}$" $1\frac{1}{2}$"

$1\frac{1}{4}$"

$\frac{3}{16}$"
BRASS

$\frac{3}{16}$" DRILL

$\frac{1}{4}$" x 1"
BRASS
BOLTS

$3\frac{1}{2}$"

$2\frac{3}{4}$"

6"

DRILL HOLES TO SUIT
CONDUIT STRAPS

FIG. 26. Shop details for a limit device for the hoist motion of a gasoline-driven crane.

and note any tendency to overturn when suddenly stopping a descending load.

On some of the smaller cranes that are subject to operation by the rank and file of employees rather than by an experienced

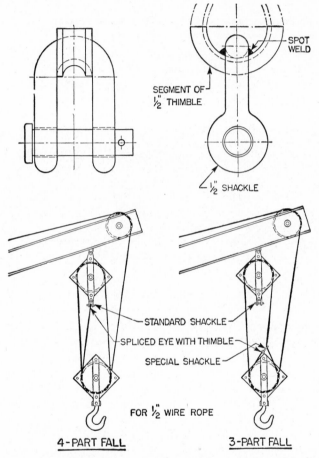

Fig. 27. A special shackle for readily changing from a three-part to four-part fall, or vice versa, with the minimum injury to the cable.

operator, limit devices are installed on the load hoist and boom hoist, and these should be tested to ensure proper operation. Figure 25 shows an indicator that can be constructed and installed by any mechanic on a boom crane to indicate directly the safe load with the boom at any angle. After the device is installed, the boom is placed at such an angle that the radius from the center pin to the load is in agreement with the data on the name plate and the safe load marked on the dial. For instance, if the name plate reads 9,500 lb at 20 ft radius, raise or lower the boom until the radius is 20 ft by actual measurement, then paint 9,500 lb on the dial.

When operating a boom crane, always make certain that the chassis is on an even keel; in other words do not have the treads or wheels on one side higher than on the other side, particularly if the load is picked up with a high boom. When the crane is out of level, there is a side bending on the boom under certain conditions, and failure to recognize this has resulted in fatal accidents.

Fig. 28. Seventy-ton automotive crane.

Do not slew the crane too rapidly while carrying a load, for not only is there danger of striking persons but there is also the possibility that the centrifugal force thus developed may upset the crane.

Crawler and Automotive Cranes. Rotary cranes mounted on automotive chassis or on crawler treads are commonly used in industry, in construction, for excavation work, and in highway building operations. These cranes are now built in capacities up to 80 tons or more (Fig. 28). When used in public streets the area in which they operate (including the swing of the crane counterweight) should be protected by conspicuously painted "horses" or other barricades of an effective type. Bright red flags should be placed on these barricades during daylight hours. At night bright red lights should outline the area. If flashing lights are used, they should have a relatively slow cycle, with the lighted period much longer than the "out" period. The neon-type flashers, with the illuminated period lasting for only a small fraction of a second, do not give effective protection.

When transporting a crawler crane on a semitrailer truck through the city streets the shortened boom in its low position extends for a considerable distance beyond the rear of the vehicle. The crane operator should remain in the cab, with the engine running, during the entire trip in order to quickly make a movement of the boom if necessary to avoid an accident.

Thoughtless car drivers frequently follow the semitrailer so closely that they actually drive under the boom. Should the wheels of the trailer strike a hole or a bump in the pavement the greatly amplified movement of the boom might cause it to crush the roof of the car. A construction vehicle, such as a pick-up truck, should always follow the trailer, beyond the end of the boom, to keep all drivers from driving beneath the boom.

A serious hazard to the crane operator and to those riggers working with him is the ever-present danger, when operating in a city street, of the boom or cables coming into contact with live overhead electric wires. The operator is not exposed to electric shock unless he attempts to leave the crane, especially if it is mounted on rubber tires. If he is holding onto the crane when his foot touches the ground he may be killed. Likewise, a rigger on the ground may be killed if he attempts to pick up a tool from the crane chassis.

Some states have labor laws requiring that all automotive cranes that operate on city streets or highways be equipped with a grounding wire which is attached to the crane chassis and to a steel plate which the crane operator lowers like an "anchor" onto the ground or pavement, even though they be perfectly dry. Those crane operators, or the riggers in charge of the job, may take safety more seriously and they may try to get a better electrical ground by connecting the wire to a fire hydrant or to a pipe driven into the earth. However, even such grounding is of questionable value, and may offer a false sense of security.

Let us assume that the voltage at the electric substation is 2,300 volts between any phase wire and ground. Any grounding wire smaller in size than the overhead wires offers no protection whatsoever, for in event of contact of the crane boom or cables with a live wire the grounding wire would be burned off before the circuit breaker (the automatic equivalent of a fuse) on the power line could open.

In Fig. 29 the grounding wire is shown to be the same size as the overhead wires, say 2/0 copper. Probably the best electrical ground available is a fire hydrant. Now, if the boom or the hoist cables of the crane make contact with a wire a current of about 1,650 amperes will flow through the crane. The voltage to ground from the crane chassis might be about 191 volts, more

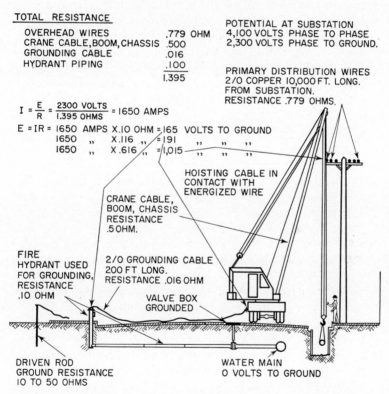

FIG. 29. Calculation of voltage at various parts of the equipment when a ground cable is attached to an automotive crane.

than enough to kill a man should he be standing on a grounded manhole cover or valve box in the pavement and touch the crane at the same time.

If no fire hydrant is available, a steel bar or pipe might be driven deep (10 ft or more) into the earth in the hope of reaching damp or wet soil.   The electrical resistance of a driven ground is quite variable, and the voltage between the crane chassis and the valve box in the pavement could be as high as 2,200 volts.   In other words, there appears to be no practicable means of adequately electrically grounding an automotive crane on a public street or highway.   Some electric public utility companies, when required to use an automotive crane within one of their substation yards, attach a grounding cable, as heavy as 500,000 cm (about 0.81-in. diameter of the copper alone), to a copper ground bus of still larger cross-section.   However, the ordinary crane owner does not have access to such a good electrical ground.

There are on the market several types of electronic devices intended to be mounted near the top of the boom.   Such devices will sound an alarm or stall the engine if brought within a predetermined distance from an energized electrical conductor. Typical alarm distances are given in Table I.

TABLE I.   SUGGESTED MINIMUM SAFE WORKING DISTANCES FROM
HIGH-VOLTAGE WIRES

Up to 4,000 volts phase to phase	5 feet
4,500 to 14,000 ,, ,, ,, ,,	10
15,000 to 27,000 ,, ,, ,, ,,	15
28,000 to 66,000 ,, ,, ,, ,,	20
67,000 to 200,000 ,, ,, ,, ,,	30

When operating an automotive crane directly below a high-tension transmission line, especially if the truck chassis and the boom are parallel to the wires, there is a possibility under certain atmospheric conditions that current from the line will leak through the air and develop a charge on the crane.   A man standing on wet ground and touching any part of the crane, the load hook, or suspended slings, or the vehicle chassis, may feel a slight shock.   However, so long as the boom and the hoisting rope are at the recommended safe distance from the wires there is no danger to the man.

All boom cranes, especially those with high booms which are used in a nearly vertical position, should be equipped with a cushioning device which will restrict the upward motion of the boom when it approaches vertical.   This will prevent the boom from being pulled too high or blown by the wind backward over the cab with disastrous results.   Figure 30 shows what can happen.

Occasionally it is necessary to gain access to a high location where work is to be done, but where ordinary means of access are inadequate. This refers to such jobs as working on a transformer on a wood line pole, or cleaning insulators or the cross arms of a steel high-tension transmission tower, or trimming a large tree, or erecting a big electric sign on the front of a building.

FIG. 30. The result of failure to provide a snubber or cushion to prevent raising a crane boom too high.

FIG. 31. Special "bucket" rig for reaching otherwise inaccessible locations.

For these kinds of jobs a special type of crane mounted on a flatbed motor truck is frequently used (Fig. 31). This outfit, which is known by various trade names such as "Snorkel," "Sky-Lift," "Sky-Master," or "Hi-Ranger," consists of a two-piece, double-hinged boom mounted upon a turntable, and provided at its upper end with a "bucket" in which one or two men may stand and work. Operation may be from either the bucket or the truck bed.

This apparatus has been used successfully in getting firemen to a strategic position at a large building fire so as to best direct the hose stream into a window or other inaccessible location. It has also been used to rescue persons trapped at upper windows and get them quickly and safely to the ground. In fact, it is standard apparatus in some large city fire departments.

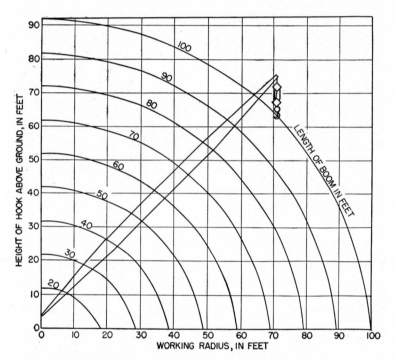

Fig. 32.   Maximum lift of load hook on a typical derrick.

Guy Derricks. To determine the safety, or the safe load, of a guy derrick it is necessary to calculate the maximum stresses imposed by the live load on the various members. The simplest way is to make a stress diagram. First we must make a scale drawing of the derrick to any convenient scale, with the boom in its lowest operating position. Use a scale, let us say, of 1 in. = 30 ft (Fig. 33a). This derrick is assumed to have a 90-ft mast and an 80-ft boom. From Table II we learn that the sheave blocks, the hook and ball, and one-half the weight of the boom may total about 5,000 lb. This figure must be added to the

20,000-lb (10-ton) live load, making 25,000-lb vertical pull on the end of the boom.

We will next draw a stress diagram (Fig. 33b) to any convenient scale, say 1 in. = 10,000 lb. (The diagram in this *Handbook* is

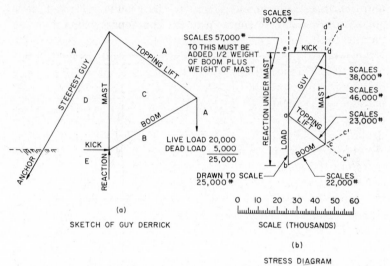

(a)

SKETCH OF GUY DERRICK

(b)

STRESS DIAGRAM

Fig. 33. Stress diagram for a typical guy derrick.

necessarily drawn to a smaller scale due to the limited size of the page.) We draw a vertical line *a-b* $2\frac{1}{2}$ in. long to represent the

TABLE II. DEAD VERTICAL LOAD ON TOPPING LIFT*

This is the approximate sum of the weights of:
a. One-half the weight of the boom
b. Weight of the upper and lower blocks
c. Weight of the overhauling ball
d. Weight of the load hook

In the case of a very high lift the weight of the rope when the hook is in its lowest position should be added.

Average loads to be used when actual weights are not obtainable.

Capacity of derrick	3 tons	5 tons	10 tons	15 tons	20 tons	30 tons
30-ft steel boom........	......	1,300 lb	1,800 lb	2,300 lb	3,300 lb	5,000 lb
40- to 50-ft steel boom.	......	1,900	2,700	3,500	4,700	6,900
55- to 75-ft steel boom.	......	2,500	3,600	4,700	6,200	8,600
Wooden boom.........	800 lb	1,600	2,000	2,600	4,000	

* American Chain and Cable Co.

25,000-lb load.   This force, shown as the live load on the sketch
of the derrick, is between the letters A and B.   Hence, the corresponding line on the stress diagram is indicated by the small
letters a-b.   This holds true for the other parts of the derrick
and their respective stresses.

Next draw a line b-c' parallel to the boom, and then a line a-c''
parallel to the topping lift A-C.   These lines cross each other at c.
Draw a line a-d' parallel to the steepest guy A-D (in the interest
of safety it is assumed that all of the stress is taken by one guy
only, which may or may not be the case), and a vertical line c-d''
parallel to the mast C-D, intersecting the other line at d.   Then
draw a vertical line b-e' to represent the downward force on the
derrick foot blocks B-E, and a horizontal line d-e'' parallel to the
ground or the roof beams on which the derrick is mounted to
represent the horizontal thrust or "kick" D-E.

Now we measure each of these lines on the stress diagram using
the same 1 in. = 10,000-lb scale, and indicate in figures the
stresses on the respective parts of the derrick.   These will be the
forces developed.   However, to b-e must be added the weight of
the other half of the boom and the weights of the mast and foot
blocks.

Using data given elsewhere in this *Handbook*, ascertain if the
proposed parts of the derrick are of adequate size or, if you prefer,
select the sizes necessary to carry the calculated stresses.

During the erection of tall steel buildings there is the problem
of periodically moving the erection derrick from one erection
floor to the next.   This job is like a man trying to lift himself by
his boot straps.   In Fig. 34a is shown a derrick in position on the
6th floor, the last of the steel having been erected to the 8th floor.
The derrick is located between the floor beams.

The boom is pulled tightly against the mast, and sufficient
strain is taken on the topping lift to permit the removal of the
hinge pin (Fig. 34b).   The boom is then rotated 180° and rested on
some timbers placed on the floor beams.   The foot of the boom is
lashed in place against accidental displacement (Fig. 34c).   Four
guys, permanently attached to the top of the boom but coiled up
and lashed to it, are secured to the building columns, and the
mast guys are slackened off (Fig. 34c).   This leaves the boom in
position as a vertical gin pole.

A wire rope sling is placed around the mast as low as practicable, but above its center of gravity, and the load hook engaged to

It (Fig. 34c).   By winding up on the load rope and simultaneously slackening the topping lift the mast is raised the 20 to 25 ft necessary to set it on the 8th floor beams (Fig. 34d).   The mast guys are again installed at the new erection floor, and the foot blocks are anchored against horizontal movement.   The boom guys are slackened off and lashed again to the boom (Fig. 34d).

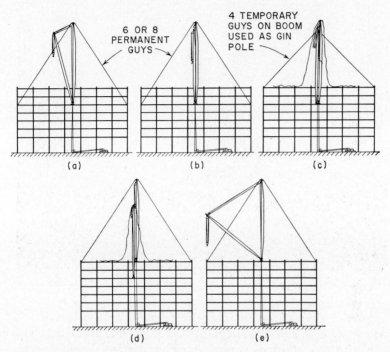

FIG. 34.   Steps in the operation of "jumping" a steel-guyed derrick from one erection floor to the next.

The boom is rotated 180° back to its original position, and the load rope slackened.   By means of the topping lift the boom is raised to its normal position and its hinge pin replaced.   The derrick is now on the 8th floor and ready for resumption of steel erection (Fig. 34c).

When installing the guys, and prior to adjusting the turnbuckles to plumb the mast, a plank is placed with its upper end inserted into the large clipped eye of the guy rope just above the turnbuckle, and the lower end placed against a column stub to prevent displacement (Fig. 35).   While the author knows of no actual occurrence, some steel companies believe that a suddenly applied load on a derrick guy (which load causes a slight unlaying of the strands of the rope) may cause a well-lubricated turnbuckle to unscrew itself and release the guy. The plank in the guy will prevent this untwisting of the guy rope. (American Bridge Co. Safety Rule 8.34)

Fig. 35. Planks are placed in the guys to prevent untwisting if a load is suddenly applied.

Special Cranes for Building Erection. For the erection of so-called "high-rise" apartment houses of reinforced concrete construction special types of cranes have been designed. Most common is the automotive type of rotary crane, as described above, but equipped with an extremely long boom having a long auxiliary or jib boom at its top (Fig. 36).

The total length of the boom may reach 250 ft or more above the street level, with the boom in a nearly vertical position. The hoist rope is usually of the 18 X 7 nonspinning type, with the live load limited to about 5 tons. Overhauling balls must be used to bring the load hook down. Of course, the crane chassis is equipped with outriggers which are jacked up so as to remove all load from the rubber tires of the vehicle.

The author feels that in order to reduce their weight many of these cranes have booms designed with a too-low factor of safety.

In fact, he believes that the height of the booms should be restricted. This is evident, judging from the frequency with which boom collapses are reported in the newspapers. Fortunately, the toll of lives has not been great. But we should realize that if a boom fell longitudinally into a city street where traffic is heavy it could conceivably crush several crowded buses within the

Fig. 36. High-lift automotive erection crane. (*Bucyrus-Erie Co.*)

block-long area, as well as private vehicles and pedestrians. (See Fig. 7 of Chapter XXI).

These cranes operate in public streets, and when the traffic barricades are placed to allow for the swing of the counterweight the usable width of the street may be reduced as much as 50 per cent. This causes traffic tie-ups for many blocks in either direction.

Even with the extremely long boom there is still a limit to the height of the building which can be erected. Also, the crane operator has to depend upon signals or walkie-talkie radio when placing his load at the erection floor.

From Europe have come what are known as "tower cranes" for erecting such buildings. One is a rotary crane known as the "traveler" type (Fig. 37), operating on a wide-gauge railroad track adjacent to the wall of a building of considerable length. The triangular boom is raised to the vertical position and acts as a

Fig. 37. Traveler type tower crane. (*Bucyrus-Erie Co.*)

mast. At its top is an auxiliary boom or jib which can be raised or lowered to change its working radius. The operator's cab is in the form of an elevator car within the boom (really a mast), and can be raised as construction progresses so that the operator always has an unobstructed view of the erection floor. This crane, also, can only be used at buildings of limited height. (See also Fig. 38.)

The "static" type of tower crane (Fig. 39) may consist of a

conventional rotary crane unit having a boom of shorter length, and mounted atop a structural steel tower erected against the wall of the building under construction. When properly guyed and secured to the building it can reach to greater heights.

Perhaps the most versatile type is the "climbing" tower crane, one make of which is shown in Fig. 40, and another in Fig. 41. Prior to the start of construction of the building, a foundation of

FIG. 38. Mobile type tower crane. (*Harnischfeger Corp.*)

four 24-in. I-beams, bolted together to form a cross, is placed on the ground, with its center located exactly at the position of one of the building elevator shafts. Heavy counterweights are placed at the outer ends of these beams. A steel tower, 4 ft 0 in. by 4 ft 0 in. by about 65 ft high, is erected on this foundation. At its top is placed a rotary crane similar to that described above for the automotive chassis, but having an 80-ft boom. (Another make has a long-reaching counterweighted boom with a trolley

FIG. 39.   Fixed or static type tower crane.   (*Harnischfeger Corp.*)

FIG. 40.   Climbing tower crane with a fixed boom.   (*Harnischfeger Corp.*)

riding on it.)   The capacity is 5 tons up to 60-ft radius, and
4 tons from 60- to 80- ft radius.

The building is erected around this tower, up to the fifth floor
above the foundation.   Inasmuch as the foundation may be at
the ground floor, or in the basement, or in the subbasement, it
will be less confusing in this text to indicate the floors by letters

Fig. 41.   Climbing tower crane with a hinged boom.   (*Climbing Tower Derrick Co.*)

rather than by the conventional numbers (Fig. 42).   As the concrete in floors *B*, *C*, and *D* becomes cured by aging, steel guide
frames are placed around the tower in the elevator shaft.   These
frames are secured to the floors.   By the time floor *E* has been
poured, floor *D* should be sufficiently cured to permit the installation of the guide frame at that level.

At floor *C* a $1\frac{1}{8}$-in. wire rope is attached to the guide frame; it
extends down the elevator shaftway along the side of the tower,
passes under a fixed sheave at the bottom of the tower, extends

upward inside the tower and over a movable sheave, down and
under another fixed sheave, and up the shaftway to floor C where
the end is anchored.   A $\frac{5}{8}$-in. wire rope from the third drum on the
hoist machine extends down the center of the tower, through the
pintle, and is reeved into a 9-part tackle attached to the movable
sheave mentioned above.   By the time all the concrete has been

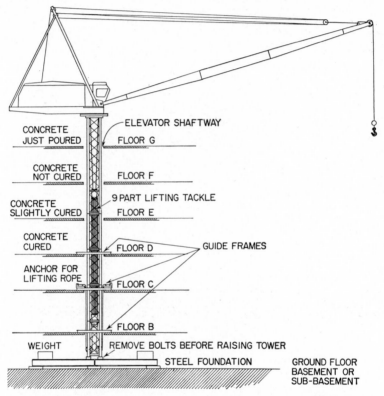

Fig. 42.   Diagram showing means of periodically raising a climbing tower
derrick.

poured on floor G it is possible to install a guide frame at floor E.
The bolts securing the base of the tower to the foundation beams
are then removed.   This leaves the tower resting on the foundation, and supported laterally by the guide frames on floors B, C,
D, and E.

By means of the hoist engine the tower is then raised "by its
bootstraps" until the lower end is just above floor B.   Hinged

50 = FOOT BOOM				
RADIUS	FT	50	35	20
CAPACITY	Tons	5	6	7
TOPPING LIFT	LB	27,000	21,000	19,000
REACTIONS IN LB	LD	39,200	32,400	21,500
	LU	38,400	31,600	20,700
	MD	80,000	74,000	61,500
	MU	24,000	13,500	—

60 = FOOT BOOM				
RADIUS	FT	60	45	30
CAPACITY	Tons	3	4	5
TOPPING LIFT	LB	21,000	19,000	20,000
REACTIONS IN LB	LD	31,200	29,600	24,000
	LU	30,400	28,800	23,200
	MD	62,500	62,800	63,100
	MU	20,000	16,000	7,800

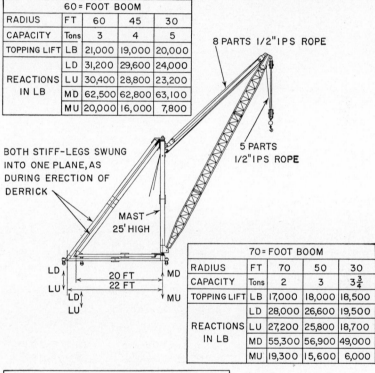

8 PARTS 1/2" IPS ROPE

5 PARTS 1/2" IPS ROPE

BOTH STIFF-LEGS SWUNG INTO ONE PLANE, AS DURING ERECTION OF DERRICK

MAST 25' HIGH

LD
LU
LD
LU

20 FT
22 FT

MD
MU

70 = FOOT BOOM				
RADIUS	FT	70	50	30
CAPACITY	Tons	2	3	$3\frac{3}{4}$
TOPPING LIFT	LB	17,000	18,000	18,500
REACTIONS IN LB	LD	28,000	26,600	19,500
	LU	27,200	25,800	18,700
	MD	55,300	56,900	49,000
	MU	19,300	15,600	6,000

80 = FOOT BOOM					
RADIUS	FT	80	65	50	35
CAPACITY	Tons	$1\frac{1}{2}$	$1\frac{3}{4}$	$2\frac{1}{4}$	$2\frac{3}{4}$
TOPPING LIFT	LB	18,000	16,000	17,000	19,000
REACTIONS IN LB	LD	27,500	24,800	22,600	18,400
	LU	26,800	24,000	21,800	17,600
	MD	54,100	50,100	51,000	47,200
	MU	20,000	16,600	13,100	7,500

NOTES: 33-1/3% ALLOWED
FOR IMPACT
IN TABLES
L INDICATES LEG
M INDICATES MAST
D INDICATES DOWNWARD
   FORCES
U INDICATES UPLIFT

FIG. 43   Typical stiff-leg derricks. (*Harlem River Truckers and Riggers, Inc.*)

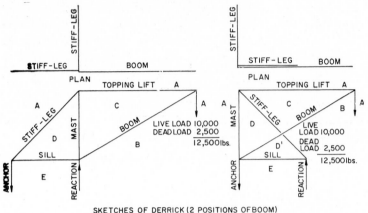

SKETCHES OF DERRICK (2 POSITIONS OF BOOM)

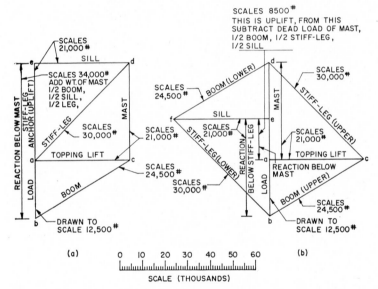

SCALE (THOUSANDS)

STRESS DIAGRAMS

Fɪɢ. 44.   Stress diagram for a stiff-leg derrick.

beams attached to the guide frames are then swung into position
and the tower lowered so as to rest on them.   Wedges are driven
in between the tower and the guide frames at all levels, and construction then starts on floor *H*.   This operation is repeated as
each floor is completed.

This type of crane has several advantages: (1) Theoretically at least there is no limit to the height of the building. (2) The public street is not obstructed by the crane. (3) The operator has an unobstructed view of the erection floor. (4) A closed-circuit television camera focused on the one spot at the street level where loads are picked up gives the operator a clear view of rigging operations at the street level.

Stiff-leg Derricks. As an example of the method of figuring the strength of a stiff-leg derrick our sketch indicates a 30-ft mast and a 60-ft boom. The load to be lifted is 10,000 lb (Fig. 44).

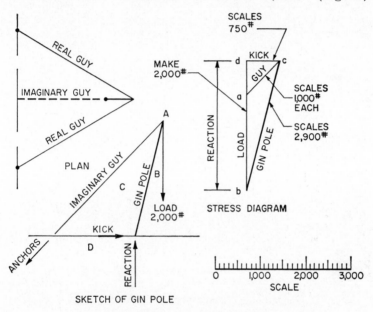

FIG. 45.   Stress diagram for a gin pole.

To determine the stresses we must make two stress diagrams, one with the boom slewed to a position opposite a stiff-leg, and one with the boom close to a stiff-leg.

After making the sketches to scale, draw the two stress diagrams in a manner similar to that for a guy derrick (Figs. 44a and 44b), using as a vertical load the live load plus allowance for one-half the boom, the hook, the blocks, etc. The stresses indicated in Fig. 44a are all greater than the corresponding members in Fig. 44b, except:

1. The stiff-leg in Fig. 44b is under compression rather than tension, but the stress has the same magnitude.

2. The vertical force below the stiff-leg in Fig. 44*b* is downward, whereas in Fig. 44*a* it is an uplift of the same amount.

3. The stress in the mast in Fig. 44*b* is greater and it is in tension.

In both cases, as with the guy derrick, the weight of one-half of the boom, the weight of the mast, and half of the two stiff-legs and

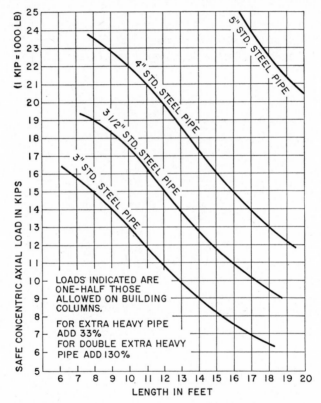

Fig. 46.   Safe axial loads on gin poles of steel pipe.   (*American Institute of Steel Construction.*)

the two sills must be added to the vertical downward force below the mast, as obtained from the diagram.   Likewise, half of the weight of the sill and the stiff-leg must be added to the vertical downward force below the stiff-leg.

Chicago Boom.   A Chicago boom is an ordinary derrick boom so installed as to utilize a building column as the mast, and to depend upon the structural steel beam connections or bracing to take the place of the stiff-legs. If heavy loads are to be lifted, steel guy ropes should be installed to carry the stress from the anchorage of the topping lift to a beam at the approximate elevation of the boom hinge or socket. The lower end of the boom is attached to the building column by means of a combination hinge pin and swivel pin to a heavy steel plate clamp attached to the column.

Gin Pole. A gin pole is merely a boom or a heavy steel pipe secured at its base against slippage and inclined at a slight angle to the vertical. Usually two guys (Figs. 14 and 45), about 60 to

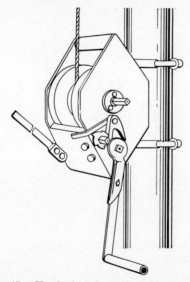

FIG. 47. Hand winch for a gin pole or tripod.

90° apart in the plan view, are attached to the top of the gin pole to take the strain due to the load to be lifted. For safety, a third (and preferably a fourth) guy is installed in front to prevent the pole from falling over backward in event of an unexpected jerk or the sudden release of the load.

Figure 45 shows the stress diagram for a gin pole, it being assumed that one of the guys takes all of the stress. When the sketch is drawn, the lower end of the imaginary single guy is shown as anchored at a line drawn between the anchorages for the real guys.

Figure 46 shows the safe axial loads on wood and pipe gin poles. For use with the gin pole is a portable hand winch (Fig. 47). This winch may also be clamped to a leg of a tripod. Figure 48 shows a portable light-weight electric winch which can be operated from an ordinary 120-volt outlet. The entire winch weighs only 35 lb.

Portable Gantry and Tripod. A handy tool for infrequent hoisting jobs in various parts of an industrial plant is a portable

Fig. 48. Portable electric winch having a 750-lb rope pull at 60 feet per minute Weighs only 30 lb and can be plugged into an ordinary 120-volt electrical recepta-cle. (*Parke Thompson Associates.*)

steel or aluminum gantry (Fig. 49). This relatively light-weight rig is mounted on lockable rubber-tired casters. The length of the I-beam is fixed, but the legs can be lengthened or shortened as desired so that if necessary a load can be lifted from the ground to a loading platform. To facilitate getting through a narrow doorway, with or without a load, the spread of the legs may be reduced at will. This rig can be dismantled for storage or, if conditions require, three of the legs can be used to make a tripod (Fig. 50).

Breast and A-frame Derricks.   A breast derrick (Fig. 13) is most commonly used by stone erectors working on large public buildings.   It is hand operated, since most lifts are only a few feet.   To calculate the stress in the guys follow the procedure for gin poles.   The A-frame derrick (Fig. 12) is most frequently built onto the deck of marine lighters.

Fig. 49.   Portable, adjustable light-duty gantry crane.   (*B. E. Wallace Corp.*)

Hoist Machines.   The hoist machine or "engine" may have two or three drums, and it may be powered by gasoline, diesel fuel, or electricity.   This machine must be properly anchored in position to prevent it from moving on the ground or floor due to the pull of the hoisting ropes.   For instance, if the maximum pull on the hoist rope is, let us say, 3,000 lb, and the pull on the topping lift rope is 2,000 lb, then the anchorage must safely withstand a total pull of 5,000 lb.   Short lengths of discarded, but usable, hoisting rope should extend from the machine bed plate to build-

ing columns or other suitable anchorages.   Lashings should also
be installed to prevent creepage sideways.

 In the case of a guy derrick erected on the ground there is a
tendency for the tension on the two ropes to pull the mast foot

Fig. 50.   Steel tripod for use in tight quarters.   Made from three legs of gantry
crane shown in Fig. 49.   (*B. E. Wallace Corp.*)

blocks and the hoist machine toward each other.   To resist this,
two heavy timbers (10 in. × 10 in. or 12 in. × 12 in.) are laid
parallel on the ground between them and secured together against
buckling by nailing planks to them.

CHAPTER XVII

HANDLING LOADS ON SLINGS

Rigging, in the truest sense, is the handling of loads suspended from a crane, hoist, or derrick. When hoisting, such objects as the top casing of a steam turbine or a large centrifugal pump must be lifted "on an even keel"; otherwise, they may foul and damage the rotor. One end of the casing may be much heavier than the other, and the eyebolts or lifting lugs may or may not have been located with respect to the center of gravity of the load.

Before attempting to lift such a load, estimate the location of its center of gravity (see Chap. I) and spot the crane hook directly over this theoretical point as shown in Fig. 1. Then procure slings of the proper strength and length to reach from the hook to the eyebolts or lifting lugs (Chap. V). This may require that one or more of the slings be of an odd length which is not readily available. If the load is lifted without regard to the position of the center of gravity (Fig. 1b, dotted lines) the load will tilt until the center of gravity is directly below the point of support, that is, the hook (Fig. 1b, solid lines). Hence, the load will be suspended at an angle.

If slings of the required length cannot be obtained, it may be necessary to substitute one or more chain hoists of adequate strength for slings. Where an odd-length sling is required, but where it is not necessary to provide the accurate leveling afforded by a chain hoist, a single sling may be rigged up as shown in the diagram in Fig. 1a and adjusted to the required length. Theoretically, this sling will slip under strain, but practice seems to indicate that friction is sufficient to hold its adjustment. However, just to play safe it is recommended that a cable clip (Fig. 10 in Chap. III) be installed where indicated on the diagram.

When placing a sling on a load, care should be used to pad all sharp corners, not merely by means of some burlap but by protecting the edges by cable guards or bent plates of adequate thickness to hold their shape under load. A good rule to follow is to make sure that the length of the arc of contact of the rope is

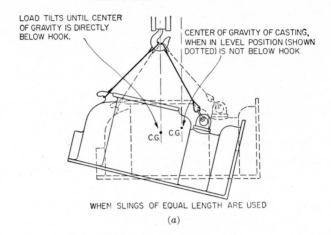

LOAD TILTS UNTIL CENTER
OF GRAVITY IS DIRECTLY
BELOW HOOK.

CENTER OF GRAVITY OF CASTING,
WHEN IN LEVEL POSITION (SHOWN
DOTTED) IS NOT BELOW HOOK.

C.G.   C.G.

WHEN SLINGS OF EQUAL LENGTH ARE USED

(*a*)

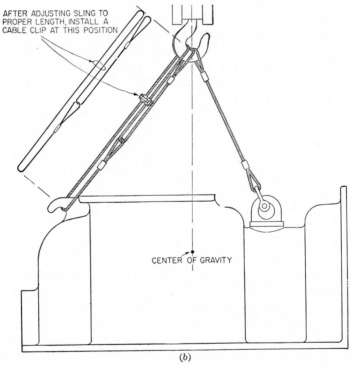

AFTER ADJUSTING SLING TO
PROPER LENGTH, INSTALL A
CABLE CLIP AT THIS POSITION

CENTER OF GRAVITY

(*b*)

Fɪɢ. 1.   Slings of unequal length are required when the center of gravity is not
midway between the hitch points.

at least equal to one rope lay (about seven times the rope diameter). When the bend is of this length, each of the strands has been on the inner and outer sides of the rope bend, and the slippage of the strands relative to each other minimizes the stress. On the other hand, if the bend is very short, the strands on the outer side of the rope bend will have to stretch, and this stretch will usually be permanent and will leave a sharp bend or "kink" in the sling.

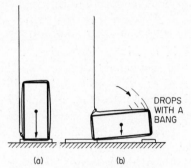

(a)   (b)

Fig. 2. Improper hitch for turning a load over onto its side.

If a load is to be rotated so as to rest on its side, instead of on its end, the sling must be properly installed; otherwise, when partially turned over, it may flop over with a bang (Fig. 2). A single sling used as a choker hitch should be arranged as shown in Fig. 3a. The part of the sling leading from the crane hook should go directly to the bottom of the load. A strain is taken

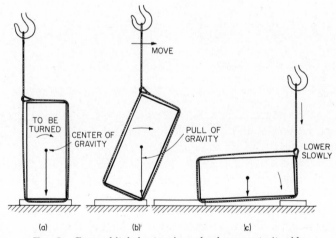

(a)   (b)   (c)

Fig. 3. Proper hitch for turning a load over onto its side.

on the sling, and the crane is moved as indicated (Fig. 3b). This causes rotation of the load; and when a vertical line downward from the center of gravity moves past the corner of the load which rests on the ground, the rotation will continue by gravity as the crane operator lowers the hook (Fig. 3c).

If the hitch point, indicated by x in Fig. 4a, is below the top corner of the load, an unsafe condition will be created.   Then when the gravity line passes the corner of the load resting on the ground the load will turn rapidly, then rock back and forth with excessive stress on the rope until the load finally comes to rest. If the hitch point is near the lower corner, it will flop over with a bang, as mentioned above.

Never should a basket sling be used in such a manner that the load can slip on it as the load is being rotated (Fig. 5), for the sharp edges will cause severe damage to the sling.   Use a choker sling instead.

To invert a heavy load, such as the upper casing of a steam turbine   or   centrifugal   pump

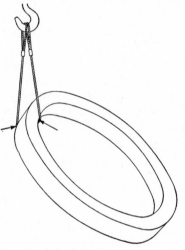

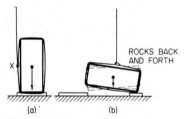

FIG. 4.   Another improper hitch for turning a load over onto its side.

FIG. 5.   A basket sling used on a lift like this will cause chafing of the sling. A choker hitch is recommended.

which has to be worked on upside down, by means of two hoists or a crane having a main hoist and an auxiliary hoist, the procedure indicated in Fig. 6 is suggested.

In order to avoid damage to the rotor, the cover must necessarily be raised in a level position.   If a single eyebolt has been provided at the top of the casing, it has probably been located directly over the center of gravity by the designer, so the lift should be made without difficulty.   The sling from the second hoist should then be attached to an eyebolt which may be located at the extreme side of the casting (the side which will be upward during the turnover) or possibly screwed into a specially tapped hole in the flange.

If there is no center eyebolt, as described above, the cover may be lifted by using two eyebolts, one high up on the casting and the other down on the flange, as shown in Fig. 6a. An imaginary line drawn through these eyebolts must pass above the center of gravity of the load (indicated by *C.G.* on the drawing). Otherwise, the load will be top-heavy and probably will roll over sideways, perhaps causing damage or personal injury.

Inasmuch as the eyebolts will be subject to pull at angles varying from a direct pull to a pull at an angle of 90 deg or more as the load is turned over, it is of vital importance that only shoulder-type eyebolts be used. Their strength is given in the

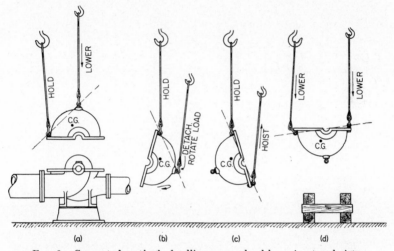

Fig. 6. Suggested method of rolling over a load by using two hoists.

lower portion of Table III in Chap. IV. Also their minimum safe strength, which is at 45 deg, should be used in the calculations. If the slings are of the bridle type instead of single slings, the weight of the casting will come onto two eyebolts instead of one during the turnover, but it is necessary to calculate the angular pull on the eyebolts.

After moving the crane bridge or trolley to a spot above a vacant floor area, the hook attached to the lower eyebolt (the main hoist hook is recommended for this position) is held in position, and the auxiliary (small) hook is carefully lowered as shown in Fig. 6b until the tilted load is suspended entirely from the big hook. The sling from the small hook is detached from the eyebolt on the top of the casting, and the load is swiveled 180 deg on the hook.

The sling is then attached to a special long-shank shoulder-type eyebolt placed through a bolt hole and held by a washer and nut, or to a shackle of adequate size and strength whose pin passes through a bolt hole in the flange (Fig. 6c). Again it must be made certain that a line drawn through the two attachment points will, when the casting is finally inverted, pass above the center of gravity of the casting.

Finally, raise the small hook so as to bring the inverted load into the desired horizontal position (Fig. 6d), and then simultaneously lower both hooks to land the load on the wood blocking

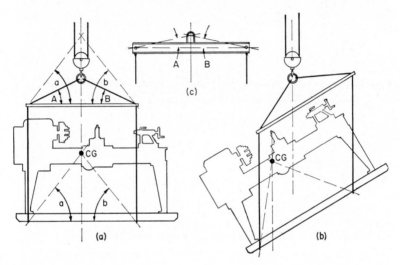

Fig. 7. Proper sling angle must be maintained when handling a load with a spreader beam.

prepared for it on the floor. Extreme care must be exercised in such maneuvers.

When using a spreader bar or rig for a 2-leg or a 4-leg sling care must be exercised to avoid spilling the load. Figure 7 shows an engine lathe mounted on wood skids for shipping. It will be noted that a lathe has a very high center of gravity, as indicated at point *CG* in sketch *a*. Draw lines on the sketch from the center of gravity to the points where the skids rest on the basket slings, these angles being *a* and *b* respectively. Then compare with the angles of the slings above the spreader. Angles *A* and *B* must be considerably greater than the respective angles *a* and *b* in order to insure stability.

Sketch *b* shows what may happen if this precaution is not taken.

for the center of gravity will tend to move to the lowest point possible directly below the point of support, which is the hoist hook. The same precautions must be taken when using a rigid spreader, as shown in sketch *c*.

To promote safety in the lifting operations of its customers one manufacturer of wire rope and slings has a team of experienced engineers conduct for their employees a 6-hour demonstration

Fig. 8. Demonstration lecture for riggers by wire-rope engineer. (*Jones & Laughlin Steel Co.*)

lecture on all phases of hoisting (Fig. 8). A motor-operated, 2-drum traveling crane of eight-foot span hoists loads, turns them over, and stands them on end as might be required at any plant. Scale models of power-plant equipment are used. This lecture and the discussion period which follows create much interest among the riggers.

(Credit for much of the material in this chapter goes to the Jones & Laughlin Steel Co. and its team of excellent instructors.)

CHAPTER XVIII

CHAIN HOISTS

No general description is needed of a chain hoist, for any rigger worthy of the title is well acquainted with it. There are, however, four types of chain hoists, namely, the spur-geared, the screw-geared, the differential, and the pull-lift types. The first three types are used for hoisting, while the last type is used primarily for pulling in a horizontal direction.

For frequent use and where a minimum of labor is available to operate it, the spur-geared hoist (Fig. 1) is recommended. Although the cost of this hoist is the highest, it will prove most economical to operate. Where the hoist is to be used infrequently, such as in a public garage, and where the first cost is a consideration, the screw-geared hoist is commonly used. For very infrequent use, such as in a private garage and where light weight and low cost are important, the differential hoist finds its place.

In the screw-geared hoist (Fig. 2) about 85 per cent of the energy the operator exerts is converted into useful work lifting the load; the other 15 per cent is wasted in overcoming friction in the gears, bearings, chains, etc. The screw-geared hoist transforms from one-third to one-half of the energy into useful work, while the differential hoist (Fig. 3) utilizes only about one-third of the energy input. Some hoist manufacturers produce a special ball-bearing differential hoist, which has a higher efficiency.

The screw-geared and differential hoists have sufficient internal friction to prevent the load from running away on the lowering motion. Such is not the case with the spur-geared hoist, so a load brake similar in principle to that described in Chap. XVI is incorporated in it.

All chain hoists are designed with their lower hooks as the weakest parts, the two hooks not being interchangeable. In other words, if the hoist is overloaded it is first indicated by the spreading or opening up of the lower hook. As designed, the inner contour of the hook is an arc of a circle, and any deviation from a

316

circle is evidence of overloading. If sufficiently overloaded, the hook will gradually straighten out (see Fig. 2 in Chap. IV) until it finally releases the load, and yet no damage should have been done to other load-bearing parts of the hoist.

 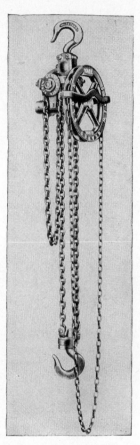

Fig. 1. Typical spur-geared chain hoist. (*Yale & Towne Mfg. Co.*)

Fig. 2. Typical screw-geared chain hoist. (*Yale & Towne Mfg. Co.*)

As shown in this illustration, a distorted hook is prima-facie evidence of not a minor overload but rather of overloading of great magnitude. All the hooks shown are rated at 1-ton capacity (2,000 lb). Even an overload of 2,000 lb (hook *a*) does not cause noticeable distortion. An overload of 3,800 lb (hook *b*) on a 2,000-lb hook causes very little spreading. Yet it is not uncommon to see chain hoists in use with the hook opened up as shown by hook *c*, and this required an overload of 4,800 lb.   Of course,

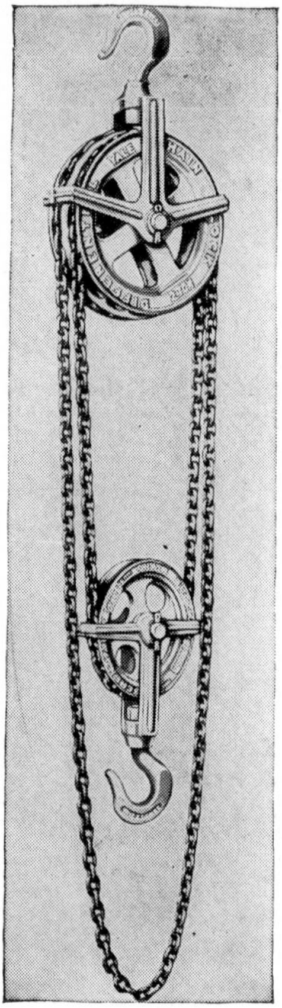

Fig. 3.   Typical differential-type chain hoist. (*Yale & Towne Mfg. Co.*)

there is no excuse whatsoever for overloading any chain hoist to this extent.

When a hook has been severely overloaded, it should be replaced by a new hook.   Never attempt to forge a spread hook back into shape.   A new hook is too cheap to warrant taking any chances with an overloaded hook.

If there is evidence of severe overloading, have the chain hoist sent to the maintenance shop for a complete internal examination and overhaul.   Pay particular attention to the wear on the brake caused by the excessive load.

Occasionally in a manufacturing plant a chain hoist is used for lowering material into oil baths or for holding material while it is sprayed with oil.   In either case the load chain may be coated with an excessive amount of oil, and when the hoist is operated the oil is transferred to the sprocket and may eventually find its way into the load brake, thus reducing its holding power.   For service such as this the screw-geared or differential hoist should be used.

Only forged steel should be used for hoist parts that are subject to stress, such as the hooks, swivels, chain, sprocket, gears, and similar parts.   Yet some hoists are made by reputable manufacturers with some such parts made of cast iron or malleable iron.

In inspecting a chain hoist it is not only necessary to examine the hooks and the general appearance of the chain carefully, but a more thorough examination is very important.   Figure 4 shows a hook of a chain hoist that was used in an industrial plant for continuous service on three shifts.   A safety inspector observed that the hook was badly worn, and he ordered it replaced by a new

TABLE I. DATA ON SPUR-GEARED CHAIN HOISTS

Rated capacity, tons	Pull on hand chain to lift rated load, lb	Hand chain overhauled to raise load 1 ft, ft
1	75	31
1½	100	35
2	110	43
3	102	70
4	112	85
5	98	128
6	115	128
8	120	170
10	125	213

hook. However, he failed to inspect the chain thoroughly. Much to the surprise of all concerned, when the hook was removed, it was discovered that the chain was in even worse condition than the hook, as indicated by the one link remaining attached to the hook.

For pulling horizontally, (Fig. 5) such as when removing tree stumps, boiler tubes, or vehicles stuck in the mud, a screw-geared

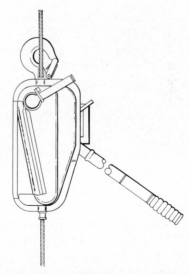

FIG. 4. This *hook* was condemned by the safety inspector. Upon removal from the chain hoist the worn link was discovered.

FIG. 5. Lever-operated device for pulling wire rope vertically or horizontally. (*Griphoist, Inc.*)

TEST GAUGES FOR CHAIN HOISTS, YALE AND TOWNE SPUR-GEARED

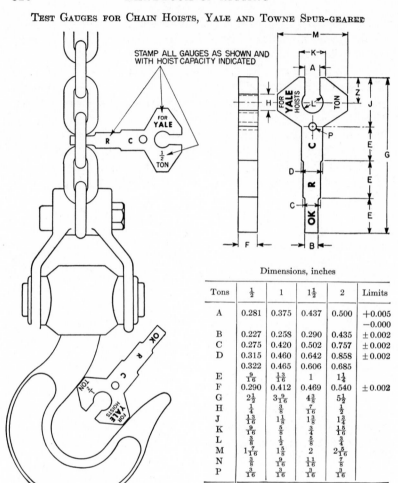

Dimensions, inches

Tons	$\frac{1}{2}$	1	$1\frac{1}{2}$	2	Limits
A	0.281	0.375	0.437	0.500	+0.005 −0.000
B	0.227	0.258	0.290	0.435	± 0.002
C	0.275	0.420	0.502	0.757	± 0.002
D	0.315	0.460	0.642	0.858	± 0.002
	0.322	0.465	0.606	0.685	
E	$\frac{9}{16}$	$\frac{13}{16}$	1	$1\frac{1}{4}$	
F	0.290	0.412	0.469	0.540	± 0.002
G	$2\frac{1}{2}$	$3\frac{9}{16}$	$4\frac{3}{8}$	$5\frac{1}{2}$	
H	$\frac{1}{4}$	$\frac{3}{8}$	$\frac{7}{16}$	$\frac{1}{2}$	
J	$\frac{13}{16}$	$1\frac{1}{8}$	$1\frac{3}{8}$	$1\frac{3}{4}$	
K	$\frac{9}{16}$	$\frac{5}{8}$	$\frac{3}{4}$	$\frac{15}{16}$	
L	$\frac{3}{8}$	$\frac{1}{2}$	$\frac{5}{8}$	$\frac{3}{4}$	
M	$1\frac{7}{16}$	$1\frac{5}{8}$	2	$2\frac{5}{16}$	
N	$\frac{3}{8}$	$\frac{9}{16}$	$\frac{11}{16}$	$\frac{7}{8}$	
P	$\frac{3}{16}$	$\frac{3}{16}$	$\frac{3}{16}$	$\frac{3}{16}$	
Mat'r'l	C. R. Steel Case Hardened				

FIG. 6. Test gauge designed by user of many chain hoists for checking the condition of the load chain. (*Consolidated Edison Co. of New York, Inc.*)

chain hoist or the special lever-operated hoist known as a pull-lift is recommended.   In this service it is almost impossible to estimate even roughly the load imposed on the hoist, and there is a possibility of overloading it.   However, if only one man pulls on the hand chain of the chain hoist, or if only one man operates the lever of the pull-lift without lengthening it by means of a piece of pipe, then there is little or no danger of overloading.

In order to develop a standard for inspecting the load chains of chain hoists, one user of a larger number of hoists has developed a set of feeler gauges shown in Fig. 6. The opening at the wide end of the gauge is used to caliper a chain link in order to make certain that the proper gauge is being used. Then, with the chain hanging freely, attempt to insert the small end of the gauge between the links as illustrated. If it is too wide, it indicates that the links have elongated and narrowed down due to overloading.

If the gauge enters the link on the first step only (stamped "OK"), it indicates little or no wear on the links. If it enters to the second step (stamped "R" to indicate "recondition"), there

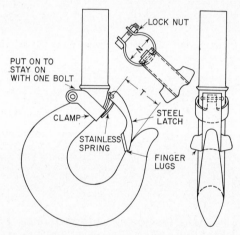

FIG. 7. A clamp-on safety finger or latch to take the place of a mousing on a hoist hook. (*The Harrington Co.*)

is evidence of considerable wear and/or stretch, and the chain hoist should be sent to the maintenance shop for reconditioning. Should the gauge enter the link to the third step (stamped "C" for "condemn"), it shows excessive wear and/or stretch and the chain should be condemned and replaced by new.

The wide end of the gauge is used to caliper the opening of the hoist hook. If the opening is in excess of the gauge, the hook should be condemned. Keep in mind that on some of the larger chain hoists on which there is a lower sheave block and consequently two or three parts of chain supporting the load, the chain will be of a smaller size than if only one part supported the load. Therefore, the size stamped on the gauge must agree with the

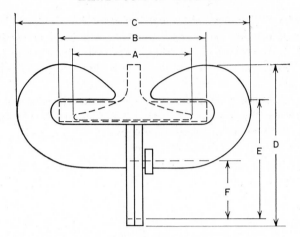

FIG. 8.   Beam clamp used to support a chain hoist from an overhead beam.
(*J. C. Renfroe & Sons, Inc.*)

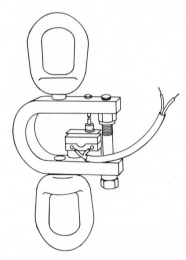

FIG. 9.   This device, attached to the hook or load block of an electric hoist, will
automatically shut off the power if an attempt is made to lift an overload.   (*W.
C. Dillon & Co.*)

capacity of the hoist when checking the hook, while a different
gauge may be required to fit the chain stock.

In Chap. XXI is described an accident in which a chain hoist
was involved.

To avoid the possibility of a sling accidentally slipping off the load hook if the load being lowered should foul an obstruction, it is good practice to install on the hook a rugged safety latch (Fig. 7). If such is not readily obtainable, make a mousing of several wraps of manila rope yarn or twine on the hook (Fig. 39, Chap. II).

For attaching a chain hoist to an overhead I-beam a special beam clamp (Fig. 8) may be used. Clamps are also available for attaching to structural channel shapes.

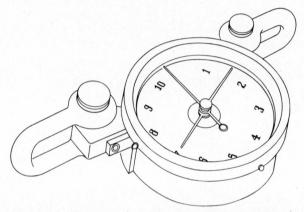

FIG. 10. Dynamometer or tension scale for weighing load being lifted. (*W. C. Dillon & Co.*)

Overloading of electric monorail hoists is a problem when these hoists are used by men who are not riggers, such as storekeepers, machine-tool operators, etc. A device such as shown in Fig. 9 can be readily attached to the load hook (and moused to it) to sound an audible alarm or cut off the power if an attempt is made to pick up an excessive load. There is also on the market a somewhat similar device which will accurately weigh the load. This is very useful for a hoist on a loading platform where the shipping weight must be determined (Fig. 10).

CHAPTER XIX

JACKS, ROLLERS, SKIDS, AND TRANSPORTATION

Jacks. There are many different styles of jacks on the market today, some of those intended for automobiles being of rather fantastic design. Such jacks should never under any circumstances be used for rigging or construction jobs. For heavy duty, hydraulic, ratchet and lever, or screw types should be used (Fig. 1).

Fɪɢ. 1. Handling a 110-ton transformer using jacks and skids. (*Consolidated Edison Co. of New York, Inc.*)

Regardless of the type of jack used, it is very important not to overload it. The lever or handle is made of a predetermined length by the manufacturer to ensure against overloading. Of course, with an extension pipe placed on the lever it is possible to lift the rated load more easily. Or with the same force on the extended lever, a greater load can be lifted. This load, however, may be in excess of the rated or safe capacity of the jack. So, do not use a longer lever than that furnished with the jack by the manufacturer.

324

In raising a load on a jack, it is important that it should not be raised so high as to run the ram or screw or ratchet out of the base and thereby drop the load. Jacks designed in accordance with the American Standard Safety Code for Jacks have "positive stops to prevent overtravel unless this is impracticable, in which case it shall carry a warning to that effect."

Always make certain that the jack is in a true vertical position (when lifting) and resting on a good footing. Never place a jack directly on the ground, even though the soil appears to be firm. If the load is to be raised in its entirety by several jacks, it should be secured laterally by struts to prevent all the jacks from upsetting in unison.

After the load has been raised to the required height, shoring or cribbing blocks should be placed under the load and wedged to take the load off the jacks or to be ready to take the load should a jack fail.

When operating a jack, particularly the ratchet and lever type, do not lean over the lever unless you are holding onto it.

Hydraulic jacks are filled with special hydraulic fluid, which will not freeze. Never use water in a jack. Keep the threads of a screw jack free from grit and dirt. Lubricate the screw frequently.

Hydraulic jacks have the safe load indicated on the name plate, but screw jacks are occasionally not labeled. Nevertheless, it is possible to overload a screw jack and cause it to fail. A very rough estimate of the safe load on a screw jack can be made by using the following formulas:

$$W = 31{,}400 \, dt \quad \text{or} \quad W = 14{,}000 \, d^2, \text{ whichever is the lesser}$$

where W = safe load on jack, lb

$\quad d$ = diameter of screw at *root* of threads, in.

$\quad t$ = thickness of nut, or length of thread engaged, in.

The load that it should be possible to lift on a screw jack should be approximately as indicated by the following formulas:

$$W = \frac{24PL}{r^2}$$

where W = load, lb

$\quad L$ = length of lever, in.

$\quad P$ = pitch of screw thread, in.

$\quad r$ = average radius of screw thread, in.

$\quad\quad$ force on lever assumed to be 120 lb

Each jack should be thoroughly inspected periodically, depending upon the service conditions. Inspections should be made not less frequently than

1. For constant or intermittent use at one locality, once every 6 months.

2. For jacks sent out of the shop for special jobs, when sent out and when returned.

3. For a jack subjected to abnormal load or shock, immediately before and after making the lift. These inspections should be made by the foreman.

Fig. 2. Unloading a 120-ton generator stator from a lighter prior to skidding it into the power plant. (*Consolidated Edison Co. of New York, Inc.*)

After a jack has been repaired and before it is returned to service, it should be tested as required by the American Standard Safety Code for Jacks, namely,

1. A load equal to the rated capacity of the jack shall be lifted to within 1 in. of its full travel under the prescribed operating conditions for the particular type of jack being tested.

2. An additional load equal to one-quarter of the rated load shall then be added to the original load applied to the jack. Any evidence that any part of the jack is stressed beyond the yield point of the material forming any part shall disqualify the jack for service.

3. When testing a jack it is important that, as the load is lifted, blocks be placed under the load to hold it in event the jack should

fail. It should be unnecessary to warn that the hands and fingers be kept from under the load during the test.

When using a jack in a horizontal position to move an object, it should be lashed or blocked up to prevent it from falling should the object unexpectedly move faster than the jack.

Rollers. For moving heavy loads across a floor or ground, hardwood (usually maple) rollers $7\frac{1}{2}$ in. diameter by 10 ft 0 in. long are commonly used (Fig. 2). Pipe rollers may also be used, but never under any condition use an oxygen cylinder or any other high-pressure gas cylinder in place of a roller. Keep in mind that these cylinders when full may have a pressure in excess of 2,000 lb per sq in.

Skids. Timber skids are commonly used under heavy machinery or other equipment that is being moved on rollers. These are, for all practical purposes, simple wooden beams the strength of which can be determined as described in Chap. XIV. Defects in the wood are discussed in Chap. VI.

Transportation. Rigging, in the broad sense, includes moving heavy loads horizontally as well as vertically. While the rigger

Fig. 3. A 238-ton, 110-ft-long boiler drum is transported to generating station. (*Consolidated Edison Co.*)

does not actually drive the vehicle, he is responsible for getting the load safely onto it, securely anchoring it in place, and removing it at its destination.

For ordinary loads a heavy motor truck usually suffices, but for heavier loads a semitrailer is necessary. Occasionally extremely heavy loads must be handled through city streets, and for these jobs specially designed semitrailers must be made available. Figure 3 shows such a rig for carrying a 150-ton section of a large power transformer. The wheel arrangement per axle on this trailer is 2—4-4—4-4—8-8.

In recent years numerous gigantic construction jobs have been undertaken by American engineers in foreign lands; many of these jobs are in desert, polar, or jungle areas. These jobs present a serious problem in the transportation of heavy equipment through regions where there are no roads or where the existing roads were not built for such large and heavy loads. Engineers and riggers responsible for work in such places may be interested

Fig. 4. Overland freight "train" having five 35-ton cars, with tires up to 10 feet in diameter and 4 feet wide, can travel over desert sands, arctic glaciers, or tropical jungle swamps. (*R. G. LeTourneau, Inc.*)

in knowing that R. G. LeTourneau, Inc., has developed a line of special equipment to meet such transportation problems, most of the equipment being designed to utilize the individual electric motor-driven wheel with a gigantic pneumatic tire.

For carrying these heavy loads an "overland freight train" (Fig. 4) consisting of a "lead car" (not a locomotive, as all vehicles are self-propelled) on which the control cab and the diesel-driven electric generator are mounted, and five "cars," each of which will carry a 35-ton load, has proved satisfactory under these very difficult conditions. This train will operate over desert sand dunes, over deep arctic snow, up 20 per cent grades on glaciers, and through tropical jungles and swamps. In the latter regions the train may be preceded by a "tree-stinger" (Fig. 5) which by brute strength uproots and fells giant trees which stand in the path. Now heavy machinery can be moved into nearly any part of the earth.

FIG. 5.   The "tree-stinger" uproots giant trees which stand in the path of the overland train.   (*R. G. LeTourneau, Inc.*)

When placing a heavy load on a semitrailer, locate it near the rear end if you want to insure stability, particularly when making turns, for the trailer has a base which is triangular in shape with the apex of the triangle at the center pin or "fifth wheel."

CHAPTER XX

HOIST SIGNALS

To reduce to the absolute minimum the number of accidents due to faulty or misunderstood signals when handling loads with cranes and hoists, it is deemed desirable to include this chapter. The signals should be thoroughly understood by the signalman and the operator, and only the approved signals should be used. It may be desirable to post a copy of the signal code in the crane cab and another copy where the signal man can occasionally refer to it.

The crane operator should take signals from no one but the authorized signalman. Where it is necessary to change signalmen frequently, they should be provided with one arm band, conspicuous hat, glove, or other "badge" of authority, which will be in the possession of the man in authority at the time. Figure 1 indicates the hand signals adopted by the ASA.

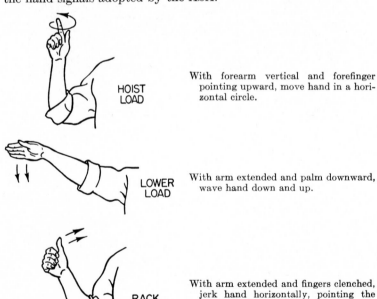

HOIST LOAD — With forearm vertical and forefinger pointing upward, move hand in a horizontal circle.

LOWER LOAD — With arm extended and palm downward, wave hand down and up.

RACK TROLLEY — With arm extended and fingers clenched, jerk hand horizontally, pointing the direction with thumb.

Fig. 1. Standard one-hand crane signals.

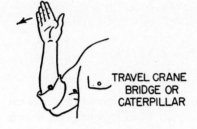

TRAVEL CRANE BRIDGE OR CATERPILLAR

With arm extended, forearm nearly vertical, and hand open with fingers pointing upward, wave hand in direction of travel while facing in that direction.

BOOM UP

With arm extended, fingers clenched, and thumb pointing upward, move hand up and down.

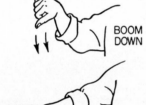

BOOM DOWN

With arm extended, fingers clenched, and thumb pointing downward, move hand down and up.

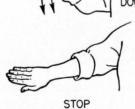

STOP

With arm extended and palm downward, hold position rigidly.

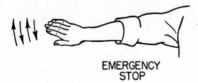

EMERGENCY STOP

With arm extended and palm downward, move hand rapidly to right and left.

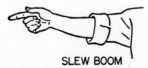

SLEW BOOM

With arm extended, point forefinger in direction of travel.

FIG. 1.   *(Continued.)*

For derricks and special temporary hoists where the hoist engineer cannot observe the signalman, bell or whistle signals may be used, different sounding bells being used for load hoist and boom hoist motions.   The signals customarily used are as follows:

HOIST SIGNALS

To hoist.....................Two quick signals
To lower.....................Three quick signals
To stop......................One quick signal
For emergency stop...........Series of quick signals

For slow or cautious motion, use slower signals and hold the last signal until it is desired to stop the motion.   Releasing the signal cord indicates "stop."

CHAPTER XXI

ACCIDENT PREVENTION

This book would not be complete without reference to recommended safe practices of riggers and the safeguarding of their equipment.

Practicing safety in his work does not bring discredit upon a worker as some people seem to think. On the contrary, if an operation is made safe, it can be done more quickly, as it will be done correctly and in an orderly fashion. Certainly no one wants to get injured, so using care will be beneficial to both the worker and the employer.

The first rule in safety is the Golden Rule: Do unto others as you would have them do unto you. Use a little forethought before you drop or throw things from an elevation to the ground by warning those below. Do not allow persons to walk below a suspended load or near a slack cable that may suddenly be pulled taut. Avoid frightening anyone intentionally or by otherwise engaging in horseplay.

Loose clothing should not be worn where it might be caught in moving machinery. Hard fiber safety helmets are generally worn by men at work where exposed to falling bolts, nuts, rivets, or other small objects. Safety shoes should be worn by *all* men on all rigging and construction jobs. This means by the foreman as well as by his men. These shoes with pressed steel caps over the toes cost no more than high-grade work shoes but are usually made of better materials and with better workmanship, and in addition they offer protection to the toes. A worker had his foot run over by a wheel of a truck, and although the wheel load was in excess of 3,000 lb, his toes were uninjured. Safety shoes can be obtained in dress oxfords as well as in high and low model work shoes, and many men wear safety shoes even for dress occasions. Of course, shoes of any type should be kept in good repair and with unbroken soles.

Where a man is exposed to the danger of falling a great distance, and where local conditions do not prohibit it, a life belt should be worn. The tail line should be as short as practicable, and it should be made fast to a substantial anchorage *above* the man's waist.   The mere wearing of a life belt just to comply with rules does not ensure safety.   An instance may be cited of a painter who was observed at work painting the underside of the high roof over a traveling crane.   A 24-ft extension ladder was placed on the bridge girder of the traveling crane, and its upper end lashed to the roof truss above at a point nearly 100 ft above the floor.   Yes, the painter wore a life belt, and he had the tail line securely anchored.   But the tail rope was not made fast to the truss member, but to the railing on the crane walkway 20 ft below him.   In other words, had he fallen, he could have dropped about 40 ft before the slack in the rope took up.   This would have meant as certain death as if no life belt had been worn.

Where there is a hazard of dust and flying particles entering the eyes, safety goggles with hardened glass lenses, or heavy plastic goggles, should be worn.   These offer protection to the eyes from small flying objects such as burrs from the mushroomed head of a cold chisel, chips from a chipping hammer, etc., and they will not shatter even though struck by a flying object of considerable weight and velocity.

Do not leave bolts, tools, slings, and other equipment lying on beams or other supports where they can become dislodged and fall onto passers-by.

The gears, couplings, and other hazardous parts of hoist engines should be protected by proper guards, and they should be kept in place at all times when the machine is in service.   Watch out for the fingers when operating dogs on ratchets of hand-operated hoists.   If the hoist is motor-operated, the frame of the machine should be electrically grounded to prevent possible electric shock or electrocution.   Gasoline engines should not exhaust their poisonous gases into confined locations or pits where men may enter unsuspectingly and be asphyxiated.

Care should be used when gassing up hoist engines.   Only an approved safety can with a self-closing cap on the spout should be used for transporting gasoline.   Gasoline vapors are heavier than air, and if gasoline is spilled, the vapors will flow like water to the lowest point, where they will accumulate.   Should a spark from a welder's torch or from any other source drop into the vapor, a flash or violent explosion may result.   Of course, gasoline should never be used for cleaning or degreasing purposes.

All injuries, however slight, should be reported, and first-aid treatment given at once in order to avoid infections and other complications at a later date.

Accident prevention is very important in all trades, but in perhaps no other trade is the danger of serious or fatal accidents so great as in rigging. Therefore, the rigger as a matter of self-preservation should be wholeheartedly interested in safety.

The description of a few unusual accidents and near accidents may be of interest. Mousings are frequently placed on hooks as a safety precaution to prevent accidental detachment of the sling, but in one instance at least a mousing was the direct cause of a serious accident. The sling was placed on the hook and the hook properly moused, but when the latter was being raised off the floor, the stiff sling rotated unnoticed and its weight rested on the mousing (Fig. 1). The mousing was strong enough to take the weight of the rope falls but failed when an additional load was applied.

FIG. 1. The upper hook of the scaffold rope fall was attached to a sling and moused (while lying on the floor). When the sling, and the pulley block, was hoisted into position the sling became displaced and rested upon the mousing, unknown to the rigger. The scaffold was hoisted into position, but when the rigger climbed a ladder and stepped onto the scaffold the mousing failed and the scaffold fell.

In another instance a 6-ton chain hoist was suspended from a sling attached to an overhead timber. As is common knowledge, chain hoists of this capacity hang on an even keel when under load, but when not loaded they tilt at a considerable angle owing to the heavy gear housing being off to one side of the upper hook. In this case a load of about 4 tons had been lifted a few inches when a part of the rigging failed and dropped the load without personal injury or property damage. But being suddenly relieved of its load, the chain hoist momentarily swung to beyond the normal tilted position and detached itself from the supporting sling and fell, narrowly missing the men below. As a safety precaution always stay out from under the hoist as well as from under the load. Figure 2 shows a typical chain hoist of 6 tons' capacity with the upper hook as it was at the time of the accident. To prevent a recurrence, swivel the hoist 180 deg relative to the hook so that it faces in the opposite direction.

This recalls to mind an accident that resulted from using two single slings at a very wide, flat angle. As shown in Fig. 3 a portion of a loading platform had to be made removable in order to permit trucks to pass by in an alley, at which time this part of the platform was placed upon the permanent platform. This meant that there was only a limited headroom under the monorail-type electric hoist. There was no means of attaching the two single slings to the platform except at its sides, and this meant having the slings at about 15 deg to the horizontal. Under normal conditions this is poor practice, but as $\frac{1}{2}$-in. wire-rope slings were available and their safe capacity at this angle was about 1,260 lb while the load was only about 1,000 lb, it was believed that the load could be safely lifted.

But the rigger failed to take one fact into consideration. The contour of the inner part of a hook is an arc of a true circle from about 45 deg above the horizontal at the back of the hook to about 45 deg below the horizontal at the side toward the bill. The one sling, pulling at 15 deg below the horizontal toward the bill of the hook, slid up the bill to the safety finger. The finger, not being designed for a force of this magnitude, sprang out of shape, the hook tilted, and the sling slipped off the bill of the hook and allowed the load to drop. So whenever it is necessary to use slings at a wide angle, use a two-leg sling with a ring on the hook rather than two single slings placed on the hook.

FIG. 2. A typical 6-ton chain hoist that became detached from the supporting sling when the load was suddenly released.

In one large plant handling bulk material, a gang of laborers was assigned the task of moving a large roll of wide conveyor belting from the storehouse to a location below a hatch in one of the conveyor bridges. Then riggers were to hoist the load up through a hatch to its destination. These laborers, who were a little above the average in ambition but a little below par in intelligence, decided to hoist the load themselves. They located some old rope that had been discarded, but not destroyed, and some pulley blocks. There was a car puller handy, so they got a hitch on the load and began to hoist. When the load was about 6 in. off the ground, one strand of the old rope snapped, and the broken ends began to unlay, so they immediately landed the load.

The laborers went into a huddle to decide what to do next, for they realized that the strand ends would not pass through the pulley blocks.   So they carefully laid the broken strand ends back into their original positions in the rope, then wrapped the rope with friction tape at this point to hold them in place, after which they proceeded to hoist the load.   Strangely, they did succeed in hoisting it about 50 ft when the other two overloaded strands failed and the load dropped.   These men should have realized that

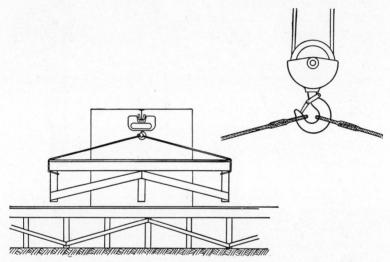

Fɪɢ. 3.   Even though strong enough to carry the load, bridle slings should not be spread at too great an angle, as one may slip off the crane or hoist hook.

if three strands would not support the load, then certainly two strands would not.   This incident demonstrates the need for always cutting all discarded rope into short lengths suitable only for hand lines, etc.

In another plant an automobile-truck crane had the load cable anchored to the boom by an ordinary wedge socket.   The crane operator, in picking up a load, carelessly allowed the load block to be raised beyond the safe limit.   The bolt of the sheave block struck the wedge and forced it back in the socket (Fig. 4), thus releasing the end of the cable, which pulled out and dropped the load.   To prevent recurrence, a cable clip was put on the two parts of the rope just ahead of the wedge.

An unusual accident occurred that resulted in minor injuries to a man working on a special lightweight 28-ft extension ladder.

Through the coincidence of the nine unsafe conditions the accident occurred (Fig. 5).

1. Dip grain existed in the side rails, and this always causes weakening of a wooden member (see Chap. VI, Fig. 18).

2. When the hardware was placed on the ladder sections, it was so installed that the dip grain was on the lower or tension side of the side rails, and dip grain is exceptionally weak in tension.

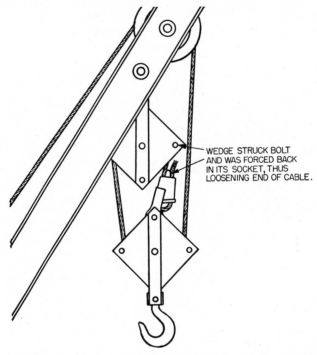

WEDGE STRUCK BOLT AND WAS FORCED BACK IN ITS SOCKET, THUS LOOSENING END OF CABLE.

Fig. 4.   Raising the load block too high caused this cable to pull out at its socket.

3. The side rails were both cut from the same plank, and thus both rails contained this defect. They were placed in the ladder in the same relative position as in the plank; thus the defects were both at the same place in the ladder.

4. The ladder builder apparently paid no attention to the dip grain, for he drilled his rung holes in both rails in the remaining good (and overstressed) wood opposite the defect.

5. In the course of its use the ladder must have been allowed to fall and strike a relatively sharp cornered object, for also in the

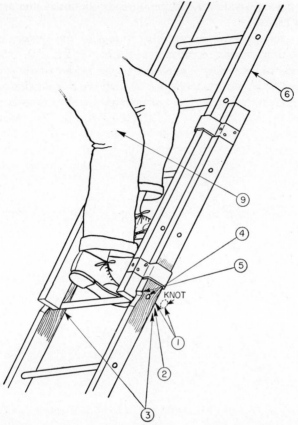

Fɪɢ. 5. Coincidence of nine conditions resulted in ladder failure.

① Dip grain, caused by a knot in the adjacent wood before sawing it up, reduced the strength of the side rail materially.

② Dip grain is particularly bad when on the tension side of a beam (the under side of a ladder side rail) as the lens-shape piece of wood tends to drop out.

③ *Both* side rails had been cut from the same plank, and placed in the ladder in their original relation to each other. Hence the dip grain caused both side rails to be weakened at the same point.

④ Ladder builder carelessly drilled the rung holes in the remaining good wood at the dip grain.

⑤ The ladder had been damaged and *both* side rails at the only remaining good wood were crushed.

⑥ Ladder was extended until bottom of fly section was at weak spot. This caused additional stress at this point.

⑦ Ladder was placed at a too-low angle, thus increasing the stress on side rails.

⑧ Man was standing on rungs at weak point on ladder.

⑨ Man was exceptionally heavy, weighing over 200 lb.

remaining good wood opposite the dip grain in both side rails were crushed or bruised spots $\frac{1}{16}$ to $\frac{1}{8}$ in. deep.

6. The defects were in the lower section of the ladder, and as the fly section was raised, its lower end was at the rung that was inserted in the weakened wood. And as is readily understood, at the upper end of the overlap of the sections of an extension ladder there is a pull of one section away from the other, while at the lower end of the overlap the sections are forced together, thus placing an additional bending load on the lower section at this point.

7. The ladder was erected at a bad angle (not steep enough) and therefore was subjected to excessive bending stress.

8. The man stood on this particular rung to work and to lift a small object.

9. The man using the ladder was extra heavy, weighing over 200 lb.

This was the man's unlucky day, for if any *one* of these conditions had not prevailed, he probably would not have been injured. It was a very unusual set of conditions, but there is no reason why similar, simultaneous conditions cannot cause accidents in the future.

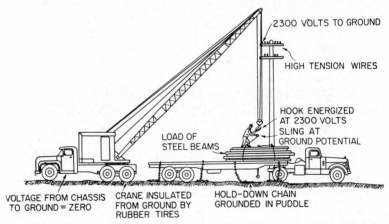

Fig. 6. This rigger completed the electrical circuit between the 2300-volt primary wire and the ground.

A truck loaded with steel beams was backed up across the sidewalk in front of a building under construction, and the driver detached a chain which had held the load in place on the truck. The end of the chain fell onto wet earth following a rain. An automotive crane on rubber tires swung its boom so as to pick up the load, but unknown to the crane operator the hoist cable came into contact with a primary (2,300 volts) power wire overhead. This energized the entire crane which was insulated from the ground by the rubber tires. The steel beams were electrically grounded by the chain resting on the wet ground. Ready to place the sling on the crane hook, the driver held the grounded sling in one hand and reached for the hook with the other

FIG. 7. Collapsed 190-ft boom of a crawler crane at a building erection job. Note the flimsy construction of the boom, especially the bracing, alternate members of which must be in compression.

hand. When he touched the hook he completed the electrical circuit and was instantly electrocuted. Watch out for live wires.

In Chapter XVI mention is made of the practice of some crane manufacturers to use dangerously low factors of safety in the design of the load-bearing members of their cranes. Figure *7* shows what remains of the 190-ft boom of a crawler crane used in building erection which failed with tragic results. Note the extremely flimsy bracing members which are subject to compressive loading.   In fact it is difficult to create a mental picture of this boom, 190 ft long, as it is being raised from its horizontal assembly position without collapsing due to its own weight.   In order to provide adequate safety for human life and valuable property, cranes should be designed with a safety factor not less than 5.   (See also Fig. 30, Chapter XVI.)

Fig. 8.   Result of overloading a suspended scaffold used by bricklayers.   Note bent outrigger beams.

The disastrous result of overloading a bricklayer's suspended scaffold is shown in Fig. 8.   The initial failure was the bending of the outrigger at the roof level.

Figure 9 shows the result of insufficient diagonal bracing of the staging, combined with excessive overloading.   The unbraced posts buckled and caused the collapse.   Also, the lack of diagonal bracing on the building side of the material hoist tower permitted the tower to be pulled over by the collapsing scaffolding.   Note how the tower posts buckled due to the lack of cross bracing.

It is fitting that a few words be said here relative to the physical strength of men. If done in a proper manner, a man can lift loads of considerable weight; yet lifting a comparatively light load improperly may result in hernia. However, with power hoisting equipment readily available as it is today, there is little need for manually lifting heavy loads. Before attempting to lift any load,

FIG. 9. Due to excessive overloading of the several stages of the scaffolding the material hoist tower was pulled over, with disastrous results.

estimate its weight and be sure that it is not too heavy for you to lift. Get your feet on a good flat surface where they will not slip and as close to the object as practicable. Then squat down, bending the knees but keeping the back straight.

Get a good firm grip on the object, and holding it close to the body, lift with the leg muscles (Fig. 10). Remember that the leg muscles are stronger than those in the back. If the object is to be moved only a very short distance, do not twist the body, for

FIG. 10.   The proper way to lift a heavy load.   (*National Safety Council.*)

the movement might result in a severe strain. Rather stand up, then step to the desired location and set down the object.

To place a load on the floor, reverse the above procedure. Keeping the back straight, bend the knees and carefully lower the object until one corner or one edge touches the floor, then carefully remove your fingers and allow the other side of the object to rest on the floor. As mentioned previously, the wearing of safety shoes is a "must" on all jobs where loads are to be lifted manually.

If the load is to be lifted say to the height of the shoulders, lift it in the manner described above and place it on a bench, table, or ledge. Then repeating the procedure, bend the knees and get a new grip on it. Keeping your back straight, straighten your knees as you lift the load.

In pushing or pulling an object, such as a suspended load, a force of 110 to 130 lb may ordinarily be exerted. In prying with a crowbar, it may be expected that a man could ordinarily push or pull about 100 lb, lift with about 200 lb, and push downward with a force not greater than his own weight. In turning a crank, such as on a winch, about 0.1 hp may be exerted by a laborer for 8 hr. For a few minutes' duration, he may even exert as much as $\frac{1}{2}$ hp.

CHAPTER XXII

CARING FOR THE INJURED

It is important in a handbook of this kind that the care of injured persons should not be overlooked. Except in isolated construction and logging camps, there are few rigging jobs where medical attention cannot be obtained in a reasonably short time. Hence, this chapter will deal only with the handling and care of persons who are critically injured until such time as the doctor or ambulance arrives.

Where jobs are in progress in isolated locations, a copy of the Red Cross first-aid manual should be kept handy for ready reference if needed.

Handling Injured Persons. The first and foremost rule is "do not move a seriously injured person unless absolutely necessary." If a man has fallen any distance, do not move him except to straighten his body out on the floor or ground in order to make him as comfortable as possible while waiting for the ambulance. Should his spine have been injured, moving him may mean instant death. In trying to get an injured man into an automobile, the sharp end of a broken rib may pierce vital organs and cause serious complications. Even moving a man with a broken leg may result in a simple fracture, which the doctor can readily set, becoming a compound fracture with the jagged end of the bone piercing and protruding through the skin.

If the patient is on the floor or ground, try to make him as comfortable as possible. Offer him a cigarette, for this will usually help him to relax. Fold up a coat and place it under his head, but do not raise it any more than necessary, and cover him with a blanket. If the ground is cold, roll up a blanket and place it close beside the patient's body. Then have three or four men all kneel on one knee (say the left knee) at the other side of the patient and slide both their hands, palms upward, under his body. At a given signal all men gently lift the patient a few inches while another man unrolls the blanket under him, and he is gently lowered onto it. The blanket is then wrapped over the patient to give the necessary warmth.

To place an injured person on a stretcher, have three or four men lift him in this manner and rest him on their knees. Another person should place the stretcher where the patient has been lying, and he is then gently lowered onto the stretcher.

To carry a severely injured person to an ambulance the "three-man carry" (Fig. 1) is used. As described in the preceding paragraph, three men kneel on one knee and slip their hands, palms upward, under the victim. Then in unison they lift his body and

THREE - MAN CARRY

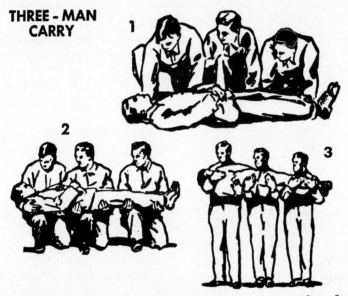

FIG. 1. The three-man carry is used for lifting a person onto a stretcher. It can also be used for carrying a person with severe back, chest, or leg injuries. (*National Safety Council.*)

rest it on their extended knees. At a signal they all stand upright and hold him close to their chests. Those carrying the patient should all walk in step, taking short uniform steps. If necessary to pass through a narrow aisle or doorway, walk side-step in unison.

Should the patient not have any broken or dislocated bones, yet be in an out-of-the-way place, it may be necessary to carry him to the ground and thence to the ambulance. If he is not too heavy, the "fireman's carry" (Fig. 2) can be used. With the patient lying face down on the floor, kneel above his head and place your arms under his armpits and grasp him at his back (Fig. 2, Step 1). Next, raise the patient up onto his knees (Step 2),

grasp him around his waist, and raise him onto his feet.   Holding
his left wrist with your right hand, "duck" your head under his
arm so that his head rests on your left shoulder (Step 3).   Still
retaining hold of his left wrist, quickly stoop and allow his body
to fall across your back at your shoulders (Step 4); at the same
time place your left arm around one or both of his legs at the knees.
Then transfer hold of his left wrist from your right to your left
hand, and stand up (Step 5).   You are then free to walk or go up
or down stairs carrying the patient and with your right hand free
to grasp handrails, etc., to assist you in your trip.

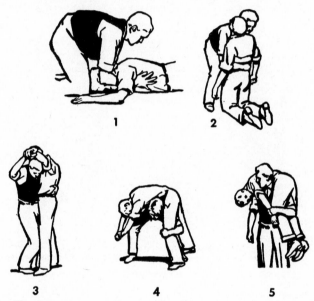

Fig. 2.   The fireman's carry allows one hand free to grasp ladder, stair railing,
etc.   (*National Safety Council.*)

In setting down the patient, the reverse procedure is followed.
Kneel on your right knee, then allow him to slide off your shoulder
and rest sitting on your extended left knee, and finally allow him
to slip down onto the floor.

For transporting a person with an injured foot or leg the "two-
man two-hand carry" is used.   An unconscious person having no
broken bones can also be carried in this manner, as shown in Fig. 3,
which is self-explanatory.

For one lone rescuer to remove an injured man from a location
with restricted headroom, lay the victim on his back and tie his

wrists together with a handkerchief or rope.   Kneel, straddling
the patient; put your head between his arms with his tied wrists
at the back of your neck; and crawl on your hands and knees,
dragging him along as you go.

To remove an injured man from a pit or a tank manhole or to
lower him from some elevated location, if no other means is available, make use of the rescue hitch shown in Fig. 85 in Chap. II.
Form a bight about 6 ft from the end of a rope, placing this bight
just below the patient's seat as he lies on the floor.   Cross the
ropes over his abdomen, then cross them again under his back
just below the shoulders, and finally pass both parts of the rope

TWO - MAN TWO - HANDED SEAT CARRY

Fig. 3.   The two-man two-handed seat carry is used for handling persons with
injured feet or ankles or uninjured unconscious persons.   (*National Safety
Council.*)

under the armpits.   Then attach the short end to the hoisting
rope some distance above his head, using a bowline.   This will
afford a reasonably comfortable sling with which a man can be
hoisted out of a manhole or other small opening.

Another method of removing an injured or unconscious man
from a manhole or other restricted space is to lash him to the
ladder that you used to reach him.   The men above will then
raise the ladder (with the victim) one rung at a time until he is
out of the manhole.

Probably the most satisfactory way of transporting an injured
person is by means of a basket-type stretcher (Fig. 4).   A person
with one or more broken bones can be strapped into the stretcher
to immobilize the injured parts of the body and thus avoid aggra-

vating the injury. When the person is thus strapped in the stretcher he may, if necessary, be handled in a vertical position; for instance, in lowering an ironworker from a building under construction to the ground, or lifting a man from a tunnel shaft.

Serious Bleeding. Look at once for serious bleeding, and if this is observed, waste no time in finding the wound from which the blood is flowing. It may be concealed by clothing. Bright red blood coming from the wound in pulsations with each beat of the heart indicates bleeding from an artery; dark red blood coming out in a steady flow is indicative of bleeding from a ruptured vein.

In all cases of arterial bleeding, pressure must be applied without a moment's delay to the artery between the wound and the heart. When bleeding is from a vein or capillary, apply pressure on a clean handkerchief (or better yet a sterile compress, if available) placed directly on the wound. Figure 5 shows the main

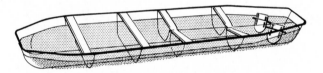

Fig. 4.   Basket-type stretcher.

arteries of the human body and the "pressure points" where pressure must be applied to arrest bleeding in the smaller arteries beyond. Study the drawing, and memorize the pressure points; you may find yourself in a position some day where your knowledge of the pressure points may save a human life.

Apply pressure at once with your thumb. If the wound is on the arm or leg, have someone apply a tourniquet while you still maintain pressure with your thumb. A tourniquet is made by stretching a handkerchief out by the diagonally opposite corners and tying it around the limb at the proper place, with a wadded-up handkerchief or other compress placed under it and directly on the artery feeding the wound. Then insert a pencil, stick, or other object in the handkerchief, and like a Spanish windless twist it so as to tighten the tourniquet and squeeze the artery. The thumb pressure may then be released.

If medical aid is delayed, loosen the tourniquet every 20 min for a few seconds to allow circulation in the limb, even though it means the loss of a little blood.   Otherwise, stagnation may cause gangrene to set in.

Resuscitation. In case of electric shock do not move the victim except to get him away from contact with the live conductor.   A victim of gas poisoning must be moved far enough to

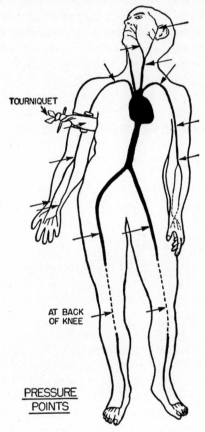

TOURNIQUET

AT BACK
OF KNEE

PRESSURE
POINTS

FIG. 5. Pressure points, or points where pressure must be applied to arteries to stop the loss of blood in arterial bleeding.

get him into fresh air.   In the case of drowning it is only necessary to get him into shallow water, or to the side of a row boat.

The Red Cross, Armed Forces, and industrial organizations now recognize mouth-to-mouth or mouth-to-nose resuscitation as the most successful and also the easiest to perform.   Resuscitation must be started immediately.   Don't waste time checking

the pulse or heart beat. Go to work at once, even if in doubt.
Every second counts; in 3 to 5 minutes it may be too late to save
the victim. A bluish tinge of the lips, tongue, and fingernails is
a danger sign.

It is most convenient to have the victim lying on his back, but
this is not necessary. Electric linemen perform resuscitation at
the top of a line pole. A drowning victim may be stood upright
in shallow water (the water makes'his body buoyant) with his
mouth above the surface. The following instructions are given
for a victim lying on his back on the ground, but they apply
equally well for other positions.

Place one hand under the victim's neck, and lift up as with
your other hand you tilt his head far backward (Fig. 6). Then
with the first hand pull his jaw forward (upward when lying

FIRST POSITION. FOR
PERFORMING MOUTH-TO-
MOUTH RESUSCITATION.
TILT THE HEAD FAR BACK

Fig. 6. Mouth-to-mouth
resuscitation, step No. 1,
tilt the head far back.
(*Edison Electric Institute.*)

SECOND POSITION. PULL
THE JAW FORWARD

Fig. 7. Step No. 2, pull
the jaw forward. (*Edison Electric Institute.*)

down) so that his tongue will not obstruct his windpipe (Fig. 7).
This alone may permit him to resume breathing.

Assuming that he does not thus resume breathing, immediately
start the resuscitation. Inhale a deep breath, open your mouth
wide, and place it over the mouth (or nose) of the victim. Use
your cheek to close and seal the nose (or mouth) so as to prevent
the escape of air, or use your fingers to close the openings. Blow
into the victim until you can observe his chest expand (Fig. 8).

Remove your mouth, and let the air escape from the victim's
respiratory system. Repeat this operation every five seconds,
or 12 times per minute. A good method of timing is to repeat
to yourself mentally "1,001, 1,002, 1,003, 1,004, 1,005."

Between breaths look into the victim's mouth for foreign solid
matter such as loose false teeth, chewing gum, etc., and remove
it with your fingers. If there is fluid in his mouth, raise his

shoulders onto your bent knee so as to have his lungs higher than his mouth. At each breath watch for the chest expansion.

Getting oxygen into the lungs, however, will not do much good if his heart is not beating and keeping a flow of blood into the lungs. So have a bystander check his pulse. This can best be felt in the arteries on either side of the neck just below the jaw-bone. If no pulse is observed, have this helper place the heel of one of his hands just below the center of the victim's breast-bone, and place his other hand on the first hand. (Fig. 9.) About once a second he should rock forward to apply pressure

THIRD POSITION. SEAL NOSE
WITH CHEEK OR WITH FINGERS,
AND BLOW INTO MOUTH

Fig. 8. Step No. 3, seal nose with cheek or with fingers, and blow into mouth. (*Edison Electric Institute.*)

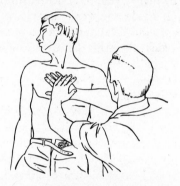

PERFORMING HEART
MESSAGE ON DEAD PERSON

Fig. 9. Step No. 4, performing heart massage on "dead" person. (*Edison Electric Institute.*)

to the breastbone. This squeezes the heart between the breast-bone and the spine, and forces blood out of the heart. Releasing the pressure allows more blood to flow into the heart.

If you are alone with a person who has no lung or heart action, you will find it possible to perform both operations. After several initial breaths, apply heart compression three or four times, breathe into the victim once or twice, apply heart pressure several times, breathe into him again, etc. Keep up both operations until there is evidence of breathing, or until a doctor takes over. Heart compression may also be used on a victim of heart attack. Some electric companies will not give up resuscitation for as long as four hours, even though a doctor has pronounced the victim dead.

While there is reluctance on the part of some people to practice mouth-to-mouth resuscitation with strangers, there is little doubt about their willingness to do so when a person's life is in jeopardy. If it is so desired, a plastic double mouthpiece is available for practice use, or a clean handkerchief may be placed between the mouths at the time of contact. Inasmuch as the operator's mouth is removed before the air is expelled by the victim there is said to be little or no danger of catching an infectious disease.

Physical Shock. Any person who has received a serious injury suffers also from physical shock. An unconscious person becomes cold very rapidly, and chilling of the body means a further strain on an already weakened vitality. Experience has shown that the cold to which unconscious persons, particularly victims of electric shock, drowning, or asphyxiation, are often carelessly exposed is probably the most important cause of pneumonia. This disease is frequently the most dangerous after-effect of even minor accidents.

As far as possible keep an injured person lying down, with the head low, while waiting for medical aid. Keep him covered with a blanket, even when performing artificial respiration. Never give an unconscious person anything to drink, as it will choke him. If he is breathing, a handkerchief wetted with aromatic spirits of ammonia may be held near but not too close to his nose. If he is conscious, hot coffee or a teaspoonful of aromatic spirits of ammonia in a small glass of water may be given the victim as a stimulant.

CHAPTER XXIII

REFERENCE CODES, LAWS, AND STANDARDS

Regardless of where hoisting or scaffolding jobs are to be undertaken, it is important to be assured that there will be no violations of local laws or, as far as practicable, any deviation from safe practices. Therefore, it is necessary for those in charge of the work to be familiar with all state and city laws and codes and with national standards and safety requirements that apply to the work to be performed.

Many of the references listed in this chapter are those which are particularly applicable in New York City. Except possibly in a very few instances the New York requirements are more stringent than those of most other cities and states. Hence, compliance with these rules should afford reasonable assurance that the work can be performed anywhere without undue criticism.

Below are listed a number of codes, laws, rules, etc., which pertain to rigging operations and which should be referred to by the riggers before starting a job. This list undoubtedly is not complete, and there are probably additional local laws that may have to be complied with.

American Standard Safety Code for Cranes, Derricks, and Hoists
  Designated ASA B30.2
  Procure from American Standards Association, New York.
    91 pages.
  Contents: Rules for design and operation of hoisting equipment.

American Standard Safety Code for Portable Wood Ladders
  Designated ASA A14.1
  Purchase from American Standards Association, New York

American Standard Safety Code for Portable Metal Ladders
  Designated ASA A14.2
  Purchase from American Standards Association, New York

American Standard Safety Code for Fixed Ladders
  Designated ASA A14.3
  Purchase from American Standards Association, New York

American Standard Safety Code for Jacks
  Designated ASA B30.1
  Purchase from American Standards Association, New York

Safety Requirements for Excavation, Building and Construction*
  Procure from Superintendent of Documents, Government
    Printing Office, Washington, D.C.
  Contents: U.S. Army regulations for the construction and use
    of scaffolds, ladders, rope, cable, chain, etc.

Wood Handbook*
  Procure from Superintendent of Documents, Government
    Printing Office, Washington, D.C.   325 pages.
  Contents: Practical handbook on the use of wood for structural
    and other purposes, prepared by the Forest Products Labora-
    tory.

*Guide Book for the Identification of Woods Used for Ties and
    Timbers**
  Designated Misc. R L2.
  Procure from Superintendent of Documents, Government
    Printing Office, Washington, D.C.   110 pages.
  Contents: As described in title.

Structural Aluminum Handbook
  Procure from Aluminum Company of America, Pittsburgh.
    106 pages.
  Contents: Properties of structural aluminum sections.

Dowmetal Extrusions
  Procure from Dow Chemical Co., Midland, Mich.   132 pages.
  Contents: Properties of structural magnesium sections.

Accident Prevention Manual for Industrial Operations
  Procure from National Safety Council, Chicago.   534 pages.
  Contents: General accident prevention, including rigging.

Modern Wire Rope Digest
  Procure from American Chain & Cable Co.   250 pages.
  Contents: Wire-rope installation design.

Safety Engineering Applied to Scaffolds
  Travelers Insurance Co. (no longer available).   350 pages.
  Contents: Data on scaffolds as of 1915.

  * All Government publications listed herein sell for a nominal price.

Handbook of Safety in Building Construction
 Travelers Insurance Co. (no longer available). 204 pages.
 Contents: Scaffolding, hoisting, etc. as of 1927.

American Recommended Practice for Inspection of Elevators
 Designated ASA A17.2
 Procure from American Standard Association, New York.
 80 pages
 Contents: Inspectors' guide

Safety Methods in Power System Construction
 Procure from Edison Electric Institute. 196 pages.
 Contents: Includes rigging.

Crane Engineering
 Procure from Whiting Corp., Harvey, Ill. 150 pages.
 Contents: Design data for traveling cranes.

Safe Practices Pamphlet No. 1—Ladders
 Procure from National Safety Council, Chicago.

Safe Practices Pamphlet No. 4—Overhead Traveling Cranes
 Procure from National Safety Council.

Safe Practices Pamphlet No. 6—Fibre Rope
 Procure from National Safety Council.

Safe Practices Pamphlet No. 12—Scaffolds
 Procure from National Safety Council.

Safe Practices Pamphlet No. 26—Wire Rope
 Procure from National Safety Council.

Safe Practices Pamphlet No. 33—Hoisting Apparatus
 Procure from National Safety Council.

Safe Practices Pamphlet No. 70—Maintenance and Repair Men
 Procure from National Safety Council.
 Contents: Life belts, slings, ladders, manual lifting.

Safe Practices Pamphlet No. 98—Use and Care of Hoisting Chain
 Procure from National Safety Council.

Crane and Hoist Engineering
 Procure from Shaw Box Crane and Hoist Division, Muskegon,
 Mich. 221 pages.
 Contents: Engineering data for design and installation of cranes.
 Mechanical details of cranes.

Roebling Handbook of Wire Rope
> Purchase from John A. Roebling's Sons Co., Trentou, N J. 240 pages.
> Contents: Installation design, maintenance, and discarding of wire rope.

Here's How
> Procure from American Cable Co., New York.   80 pages.
> Contents: Wire-rope installations.

Mechanical Engineers' Handbook, by Lionel S. Marks
> Purchase from McGraw-Hill Book Co., New York.
> Contents: Complete reference for engineers.

Rope Knots and Climbing, Bulletin No. 7, by National Park Service.
> Purchase from Superintendent of Documents, Washington, D.C.
> Contents: Use of manila rope in forestry.

Safety in Lifting
> Procure from Jones & Laughlin Steel Co.
> Contents: Complete text on the proper use of slings, etc.

Review of Structural Design for Professional Engineers License by S. W. Spielvogel
> Contents: Basic structural design.

Resuscitation Manual
> Procure from Edison Electric Institute, New York.
> Contents: Resuscitation instruction.

Steel Construction Manual
> Procure from American Institute of Steel Construction, New York.   420 pages.
> Contents: Tables of properties of structural shapes, rivets, bolts, etc.

Manual of Accident Prevention in Construction
> Procure from Associated General Contractors of America, Washington, D.C.   370 pages.

Best's Safety Directory
> Purchase from Alfred M. Best Co., Inc., New York.
> Contents: Condensed catalogues of safety and fire protection equipment.

APPENDIX

HANDY REFERENCE TABLES

TRIGONOMETRIC FUNCTIONS

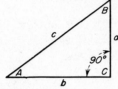 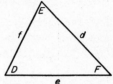

Given	Required	Formulas
a, c	A, B, b	$\sin A = \dfrac{a}{c};$ $\qquad\qquad \cos B = \dfrac{a}{c};$ $\qquad b = \sqrt{c^2 - a^2}$
a, c	Area	$\dfrac{a}{2}\sqrt{c^2 - a^2}$
a, b	A, B, c	$\tan A = \dfrac{a}{b};$ $\qquad\qquad \tan B = \dfrac{b}{a};$ $\qquad c = \sqrt{a^2 + b^2}$
a, b	Area	$\dfrac{ab}{2}$
A, a	B, b, c	$B = 90° - A;$ $\qquad\quad b = a\cot A;$ $\qquad c = \dfrac{a}{\sin A}$
A, a	Area	$\dfrac{a^2 \cot A}{2}$
A, b	B, a, c	$B = 90° - A;$ $\qquad\quad a = b\tan A;$ $\qquad c = \dfrac{b}{\cos A}$
A, b	Area	$\dfrac{b^2 \tan A}{2}$
A, c	B, a, b	$B = 90° - A;$ $\qquad\quad a = c\sin A;$ $\qquad b = c\cos A$
A, c	Area	$\dfrac{c^2 \sin 2A}{4}$
d, e, f	D	$\sin \dfrac{D}{2} = \sqrt{\dfrac{(s-e)(s-f)}{ef}};$ $\quad s = \dfrac{d+e+f}{2}$
d, e, f	E	$\sin \dfrac{E}{2} = \sqrt{\dfrac{(s-d)(s-f)}{df}}$
d, e, f	F	$\sin \dfrac{F}{2} = \sqrt{\dfrac{(s-d)(s-e)}{de}}$

TRIGONOMETRIC FUNCTIONS (*Continued*)

Given	Required	Formulas
d, e, f	Area	$\sqrt{s(s-d)(s-e)(s-f)}$
d, D, E	e, f	$e = \dfrac{d \sin E}{\sin D}; \qquad f = \dfrac{d \sin F}{\sin D}$
d, D, E	Area	$\dfrac{de \sin F}{2}$ (*e* from above formula)
d, e, D	E	$\sin E = \dfrac{e \sin D}{d}$
d, e, D	f	$f = \dfrac{e \sin F}{\sin E}$
d, e, F	D	$\tan D = \dfrac{d \sin F}{e - d \cos F}$
d, e, F	f	$f = \dfrac{d \sin F}{\sin D}$ (*D* from above formula)
d, e, F	Area	$\dfrac{de \sin F}{2}$

TRIGONOMETRIC SOLUTIONS OF RIGHT TRIANGLES

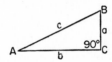

Known	Required				
	A	B	a	b	c
a, b	$\tan A = \dfrac{a}{b}$	$\cot B = \dfrac{a}{b}$			$\sqrt{a^2 + b^2}$
a, c	$\sin A = \dfrac{a}{c}$	$\cos B = \dfrac{a}{c}$		$\sqrt{c^2 - a^2}$	
b, c	$\cos A = \dfrac{b}{c}$	$\sin B = \dfrac{b}{c}$	$\sqrt{c^2 - b^2}$		
A, a		$90° - A$		$a \cot A$	$\dfrac{a}{\sin A}$
A, b		$90° - A$	$b \tan A$		$\dfrac{b}{\cos A}$
A, c		$90° - A$	$c \sin A$	$c \cos A$	

AREAS OF PLANE FIGURES

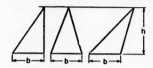

Triangle: Area $= \dfrac{bh}{2}$

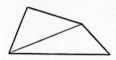

Trapezium: Area = Sum of areas of component triangles

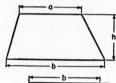

Trapezoid: Area $= \dfrac{(a + b)h}{2}$

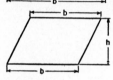

Parallelogram: Area $= bh$

Regular polygon: Area $= \dfrac{d \times \text{perimeter}}{4}$

Circle: Area $= 0.7854d^2$

Segment: Area $= \dfrac{r^2}{2}(0.0175A - \sin A)$

Sector: Area $= 0.00873r^2A = \dfrac{Pr}{2}$

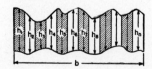

Irregular shape: divide length into parallel strips of equal width.

Area $= b\dfrac{h_1 + h_2 + h_3 + \cdots + h_n}{n}$ (approx.)

VOLUMES OF SOLID FIGURES

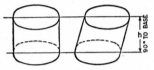

Regular prism: Volume = area of base $\times h$

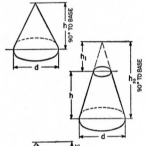

Cylinder: Volume = area of base $\times h$

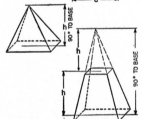

Cone: Volume = $\dfrac{\text{area of base} \times h}{3}$

Frustum of cone: Volume = volume of cone of height h_2 — volume of cone of height h_1

Pyramid: Volume = $\dfrac{\text{area of base} \times h}{3}$

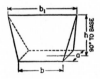

Frustum of pyramid: Volume = volume of pyramid of height h_2 — volume of cone of height h_1

Sphere: Volume = $0.524d^3$

Wedge: (Rectangular base, b_1 parallel to b)

    Volume = $\dfrac{ha(2b + b_1)}{6}$

Ring: Volume = $2.46Dd^2$

Barrel: Volume = $0.262h(2D^2 + d^2)$

CIRCUMFERENCES AND AREAS OF CIRCLES *

Diameter	Circumference	Area	Diameter	Circumference	Area	Diameter	Circumference	Area	Diameter	Circumference	Area
			7/8	2.749	0.6013	4	12.57	12.57	9	28.27	63.62
1/64	0.04909	.00019	57/64	2.798	0.6230	1/16	12.76	12.96	1/8	28.67	65.40
1/32	0.09817	.00077	29/32	2.847	0.6450	1/8	12.96	13.36	1/4	29.06	67.20
3/64	0.1473	.00173	59/64	2.896	0.6675	3/16	13.16	13.77	3/8	29.45	69.03
1/16	0.1963	.00307	15/16	2.945	0.6903	1/4	13.35	14.19	1/2	29.85	70.88
5/64	0.2454	.00479	61/64	2.994	0.7135	5/16	13.55	14.61	5/8	30.24	72.76
3/32	0.2945	.00690	31/32	3.043	0.7371	3/8	13.74	15.03	3/4	30.63	74.66
7/64	0.3436	.00940	63/64	3.093	0.7610	7/16	13.94	15.47	7/8	31.02	76.59
1/8	0.3927	.01227	1	3.142	0.7854	1/2	14.14	15.90	10	31.42	78.54
9/64	0.4418	.01553	1 1/16	3.338	0.8866	9/16	14.33	16.35	1/8	31.81	80.52
5/32	0.4909	.01917	1 1/8	3.534	0.9940	5/8	14.53	16.80	1/4	32.20	82.52
11/64	0.5400	.02320	1 3/16	3.731	1.108	11/16	14.73	17.26	3/8	32.59	84.54
3/16	0.5890	.02761	1/4	3.927	1.227	3/4	14.92	17.72	1/2	32.99	86.59
13/64	0.6381	.03241	5/16	4.123	1.353	13/16	15.12	18.19	5/8	33.38	88.66
7/32	0.6872	.03758	3/8	4.320	1.485	7/8	15.32	18.67	3/4	33.77	90.76
15/64	0.7363	.04314	7/16	4.516	1.623	15/16	15.51	19.15	7/8	34.16	92.89
1/4	0.7854	.04909	1/2	4.712	1.767	5	15.71	19.63	11	34.56	95.03
17/64	0.8345	.05542	9/16	4.909	1.917	1/16	15.90	20.13	1/8	34.95	97.21
9/32	0.8836	.06213	5/8	5.105	2.074	1/8	16.10	20.63	1/4	35.34	99.40
19/64	0.9327	.06922	11/16	5.301	2.237	3/16	16.30	21.14	3/8	35.74	101.6
5/16	0.9817	.07670	3/4	5.498	2.405	1/4	16.49	21.65	1/2	36.13	103.9
21/64	1.031	.08456	13/16	5.694	2.580	5/16	16.69	22.17	5/8	36.52	106.1
11/32	1.080	.09281	7/8	5.890	2.761	3/8	16.89	22.69	3/4	36.91	108.4
23/64	1.129	.1014	15/16	6.087	2.948	7/16	17.08	23.22	7/8	37.31	110.8
3/8	1.178	.1104	2	6.283	3.142	1/2	17.28	23.76	12	37.70	113.1
25/64	1.227	.1198	1/16	6.480	3.341	9/16	17.48	24.30	1/8	38.09	115.5
13/32	1.276	.1296	1/8	6.676	3.547	5/8	17.67	24.85	1/4	38.48	117.9
27/64	1.325	.1398	3/16	6.872	3.758	11/16	17.87	25.41	3/8	38.88	120.3
7/16	1.374	.1503	1/4	7.069	3.976	3/4	18.06	25.97	1/2	39.27	122.7
29/64	1.424	.1613	5/16	7.265	4.200	13/16	18.26	26.53	5/8	39.66	125.2
15/32	1.473	.1726	3/8	7.461	4.430	7/8	18.46	27.11	3/4	40.06	127.7
31/64	1.522	.1843	7/16	7.658	4.666	15/16	18.65	27.69	7/8	40.45	130.2
1/2	1.571	.1963	1/2	7.854	4.909	6	18.85	28.27	13	40.84	132.7
33/64	1.620	.2088	9/16	8.050	5.157	1/8	19.24	29.46	1/8	41.23	135.3
17/32	1.669	.2217	5/8	8.247	5.412	1/4	19.63	30.68	1/4	41.63	137.9
35/64	1.718	.2349	11/16	8.443	5.673	3/8	20.03	31.92	3/8	42.02	140.5
9/16	1.767	.2485	3/4	8.639	5.940	1/2	20.42	33.18	1/2	42.41	143.1
37/64	1.816	.2625	13/16	8.836	6.213	5/8	20.81	34.47	5/8	42.80	145.8
19/32	1.865	.2769	7/8	9.032	6.492	3/4	21.21	35.78	3/4	43.20	148.5
39/64	1.914	.2916	15/16	9.228	6.777	7/8	21.60	37.12	7/8	43.59	151.2
5/8	1.963	.3068	3	9.425	7.069	7	21.99	38.48	14	43.98	153.9
41/64	2.013	.3223	1/16	9.621	7.366	1/8	22.38	39.87	1/8	44.37	156.7
21/32	2.062	.3382	1/8	9.817	7.670	1/4	22.78	41.28	1/4	44.77	159.5
43/64	2.111	.3545	3/16	10.01	7.980	3/8	23.17	42.72	3/8	45.16	162.3
11/16	2.160	.3712	1/4	10.21	8.296	1/2	23.56	44.18	1/2	45.55	165.1
45/64	2.209	.3883	5/16	10.41	8.618	5/8	23.95	45.66	5/8	45.95	168.0
23/32	2.258	.4057	3/8	10.60	8.946	3/4	24.35	47.17	3/4	46.34	170.9
47/64	2.307	.4236	7/16	10.80	9.281	7/8	24.74	48.71	7/8	46.73	173.8
3/4	2.356	.4418	1/2	11.00	9.621	8	25.13	50.27	15	47.12	176.7
49/64	2.405	.4604	9/16	11.19	9.968	1/8	25.53	51.85	1/8	47.52	179.7
25/32	2.454	.4794	5/8	11.39	10.32	1/4	25.92	53.46	1/4	47.91	182.7
51/64	2.503	.4987	11/16	11.58	10.68	3/8	26.31	55.09	3/8	48.30	185.7
13/16	2.553	.5185	3/4	11.78	11.04	1/2	26.70	56.75	1/2	48.69	188.7
53/64	2.602	.5386	13/16	11.98	11.42	5/8	27.10	58.43	5/8	49.09	191.7
27/32	2.651	.5591	7/8	12.17	11.79	3/4	27.49	60.13	3/4	49.48	194.8
55/64	2.700	.5800	15/16	12.37	12.18	7/8	27.88	61.86	7/8	49.87	197.9

* From Marks, *Mechanical Engineers' Handbook*, 4th ed., McGraw-Hill, 1941.

CIRCUMFERENCES AND AREAS OF CIRCLES (*Continued*)

Diameter	Circum-ference	Area	Diameter	Circum-ference	Area	Diameter	Circum-ference	Area	Diameter	Circum-ference	Area
16	50.27	201.1	19 1/2	61.26	298.6	23	72.26	415.5	29	91.11	660.5
1/8	50.66	204.2	5/8	61.65	302.5	1/8	72.65	420.0	1/4	91.89	672.0
1/4	51.05	207.4	3/4	62.05	306.4	1/4	73.04	424.6	1/2	92.68	683.5
3/8	51.44	210.6	7/8	62.44	310.2	3/8	73.43	429.1	3/4	93.46	695.1
1/2	51.84	213.8	20	62.83	314.2	1/2	73.83	433.7	30	94.25	706.9
5/8	52.23	217.1	1/8	63.22	318.1	5/8	74.22	438.4	1/4	95.03	718.7
3/4	52.62	220.4	1/4	63.62	322.1	3/4	74.61	443.0	1/2	95.82	730.6
7/8	53.01	223.7	3/8	64.01	326.1	7/8	75.01	447.7	3/4	96.60	742.6
17	53.41	227.0	1/2	64.40	330.1	24	75.40	452.4	31	97.39	754.8
1/8	53.80	230.3	5/8	64.80	334.1	1/4	76.18	461.9	1/4	98.17	767.0
1/4	54.19	233.7	3/4	65.19	338.2	1/2	76.97	471.4	1/2	98.96	779.3
3/8	54.59	237.1	7/8	65.58	342.2	3/4	77.75	481.1	3/4	99.75	791.7
1/2	54.98	240.5	21	65.97	346.4	25	78.54	490.9	32	100.5	804.2
5/8	55.37	244.0	1/8	66.37	350.5	1/4	79.33	500.7	1/4	101.3	816.9
3/4	55.76	247.4	1/4	66.76	354.7	1/2	80.11	510.7	1/2	102.1	829.6
7/8	56.16	250.9	3/8	67.15	358.8	3/4	80.90	520.8	3/4	102.9	842.4
18	56.55	254.5	1/2	67.54	363.1	26	81.68	530.9	33	103.7	855.3
1/8	56.94	258.0	5/8	67.94	367.3	1/4	82.47	541.2	1/4	104.5	868.3
1/4	57.33	261.6	3/4	68.33	371.5	1/2	83.25	551.5	1/2	105.2	881.4
3/8	57.73	265.2	7/8	68.72	375.8	3/4	84.04	562.0	3/4	106.0	894.6
1/2	58.12	268.8	22	69.12	380.1	27	84.82	572.6	34	106.8	907.9
5/8	58.51	272.4	1/8	69.51	384.5	1/4	85.61	583.2	1/4	107.6	921.3
3/4	58.90	276.1	1/4	69.90	388.8	1/2	86.39	594.0	1/2	108.4	934.8
7/8	59.30	279.8	3/8	70.29	393.2	3/4	87.18	604.8	3/4	109.2	948.4
19	59.69	283.5	1/2	70.69	397.6	28	87.96	615.8	35	110.0	962.1
1/8	60.08	287.3	5/8	71.08	402.0	1/4	88.75	626.8	1/4	110.7	975.9
1/4	60.48	291.0	3/4	71.47	406.5	1/2	89.54	637.9	1/2	111.5	989.8
3/8	60.87	294.8	7/8	71.86	411.0	3/4	90.32	649.2	3/4	112.3	1003.8

AREAS OF CIRCLES*

DIAMETERS IN FEET AND INCHES, AREAS IN SQUARE FEET

Feet	Inches											
	0	1	2	3	4	5	6	7	8	9	10	11
0	.0000	.0055	.0218	.0491	.0873	.1364	.1963	.2673	.3491	.4418	.5454	.6600
1	.7854	.9218	1.069	1.227	1.396	1.576	1.767	1.969	2.182	2.405	2.640	2.885
2	3.142	3.409	3.687	3.976	4.276	4.587	4.909	5.241	5.585	5.940	6.305	6.681
3	7.069	7.467	7.876	8.296	8.727	9.168	9.621	10.08	10.56	11.04	11.54	12.05
4	12.57	13.10	13.64	14.19	14.75	15.32	15.90	16.50	17.10	17.72	18.35	18.99
5	19.63	20.29	20.97	21.65	22.34	23.04	23.76	24.48	25.22	25.97	26.73	27.49
6	28.27	29.07	29.87	30.68	31.50	32.34	33.18	34.04	34.91	35.78	36.67	37.57
7	38.48	39.41	40.34	41.28	42.24	43.20	44.18	45.17	46.16	47.17	48.19	49.22
8	50.27	51.32	52.38	53.46	54.54	55.64	56.75	57.86	58.99	60.13	61.28	62.44
9	63.62	64.80	66.00	67.20	68.42	69.64	70.88	72.13	73.39	74.66	75.94	77.24
10	78.54	79.85	81.18	82.52	83.86	85.22	86.59	87.97	89.36	90.76	92.18	93.60
11	95.03	96.48	97.93	99.40	100.9	102.4	103.9	105.4	106.9	108.4	110.0	111.5
12	113.1	114.7	116.3	117.9	119.5	121.1	122.7	124.4	126.0	127.7	129.4	131.0
13	132.7	134.4	136.2	137.9	139.6	141.4	143.1	144.9	146.7	148.5	150.3	152.1
14	153.9	155.8	157.6	159.5	161.4	163.2	165.1	167.0	168.9	170.9	172.8	174.8

FROM INCHES AND FRACTIONS OF AN INCH TO DECIMALS OF A FOOT*

Inches	1	2	3	4	5	6	7	8	9	10	11
Feet	0.0833	0.1667	0.2500	0.3333	0.4167	0.5000	0.5833	0 6667	0.7500	0.8333	0.9167
Inches	1/8	1/4	3/8	1/2	5/8	3/4	7/8				
Feet	0.0104	0.0208	0.0313	0.0417	0.0521	0.0625	0.0729				

EXAMPLE: 5 ft 7 3/8 in. = 5.0 + 0.5833 + 0.0313 = 5.6146 ft

* From Marks, *Mechanical Engineers' Handbook*, 4th ed., McGraw-Hill, 1941.

NATURAL SINES AND COSINES *

Natural Sines at intervals of 0°.1, or 6′.

Degrees	°.0 = (0′)	°.1 (6′)	°.2 (12′)	°.3 (18′)	°.4 (24′)	°.5 (30′)	°.6 (36′)	°.7 (42′)	°.8 (48′)	°.9 (54′)			Average difference
											0.0000	90°	
0°	0.0000	0017	0035	0052	0070	0087	0105	0122	0140	0157	0175	89	17
1	0175	0192	0209	0227	0244	0262	0279	0297	0314	0332	0349	88	17
2	0349	0366	0684	0401	0419	0436	0454	0471	0488	0506	0523	87	17
3	0523	0541	0558	0576	0593	0610	0628	0645	0663	0680	0698	86	17
4	0698	0715	0732	0750	0767	0785	0802	0819	0837	0854	0.0872	85	17
5	0.0872	0889	0906	0924	0941	0958	0976	0993	1011	1028	1045	84	17
6	1045	1063	1080	1097	1115	1132	1149	1167	1184	1201	1219	83	17
7	1219	1236	1253	1271	1288	1305	1323	1340	1357	1374	1392	82	17
8	1392	1409	1426	1444	1461	1478	1495	1513	1530	1547	1564	81	17
9	1564	1582	1599	1616	1633	1650	1668	1685	1702	1719	0.1736	80°	17
10°	0.1736	1754	1771	1788	1805	1822	1840	1857	1874	1891	1908	79	17
11	1908	1925	1942	1959	1977	1994	2011	2028	2045	2062	2079	78	17
12	2079	2096	2113	2130	2147	2164	2181	2198	2215	2233	2250	77	17
13	2250	2267	2284	2300	2317	2334	2351	2368	2385	2402	2419	76	17
14	2419	2436	2453	2470	2487	2504	2521	2538	2554	2571	0.2588	75	17
15	0.2588	2605	2622	2639	2656	2672	2689	2706	2723	2740	2756	74	17
16	2756	2773	2790	2807	2823	2840	2857	2874	2890	2907	2924	73	17
17	2924	2940	2957	2974	2990	3007	3024	3040	3057	3074	3090	72	17
18	3090	3107	3123	3140	3156	3173	3190	3206	3223	3239	3256	71	17
19	3256	3272	3289	3305	3322	3338	3355	3371	3387	3404	0.3420	70°	16
20°	0.3420	3437	3453	3469	3486	3502	3518	3535	3551	3567	3584	69	16
21	3584	3600	3616	3633	3649	3665	3681	3697	3714	3730	3746	68	16
22	3746	3762	3778	3795	3811	3827	3843	3859	3875	3891	3907	67	16
23	3907	3923	3939	3955	3971	3987	4003	4019	4035	4051	4067	66	16
24	4067	4083	4099	4115	4131	4147	4163	4179	4195	4210	0.4226	65	16
25	0.4226	4242	4258	4274	4289	4305	4321	4337	4352	4368	4384	64	16
26	4384	4399	4415	4431	4446	4462	4478	4493	4509	4524	4540	63	16
27	4540	4555	4571	4586	4602	4617	4633	4648	4664	4679	4695	62	16
28	4695	4710	4726	4741	4756	4772	4787	4802	4818	4833	4848	61	15
29	4848	4863	4879	4894	4909	4924	4939	4955	4970	4985	0.5000	60°	15
30°	0.5000	5015	5030	5045	5060	5075	5090	5105	5120	5135	5150	59	15
31	5150	5165	5180	5195	5210	5225	5240	5255	5270	5284	5299	58	15
32	5299	5314	5329	5344	5358	5373	5388	5402	5417	5432	5446	57	15
33	5446	5461	5476	5490	5505	5519	5534	5548	5563	5577	5592	56	15
34	5592	5606	5621	5635	5650	5664	5678	5693	5707	5721	0.5736	55	14
35	0.5736	5750	5764	5779	5793	5807	5821	5835	5850	5864	5878	54	14
36	5878	5892	5906	5920	5934	5948	5962	5976	5990	6004	6018	53	14
37	6018	6032	6046	6060	6074	6088	6101	6115	6129	6143	6157	52	14
38	6157	6170	6184	6198	6211	6225	6239	6252	6266	6280	6293	51	14
39	6293	6307	6320	6334	6347	6361	6374	6388	6401	6414	0.6428	50°	13
40°	0.6428	6441	6455	6468	6481	6494	6508	6521	6534	6547	6561	49	13
41	6561	6574	6587	6600	6613	6626	6639	6652	6665	6678	6691	48	13
42	6691	6704	6717	6730	6743	6756	6769	6782	6794	6807	6820	47	13
43	6820	6833	6845	6858	6871	6884	6896	6909	6921	6934	6947	46	13
44	6947	6959	6972	6984	6997	7009	7022	7034	7046	7059	0.7071	45°	12
45°	0.7071												

	°.9 = (54′)	°.8 (48′)	°.7 (42′)	°.6 (36′)	°.5 (30′)	°.4 (24′)	°.3 (18′)	°.2 (12′)	°.1 (6′)	°.0 (0′)	Degrees	

Natural Cosines

* From Marks, *Mechanical Engineers' Handbook*, 4th ed., McGraw-Hill, 1941.

NATURAL SINES AND COSINES (Continued)

Natural Sines at intervals of 0°.1, or 6'.

Degrees	°.0 =(0')	°.1 (6')	°.2 (12')	°.3 (18')	°.4 (24')	°.5 (30')	°.6 (36')	°.7 (42')	°.8 (48')	°.9 (54')			Average difference
											0.7071	45°	
45°	0.7071	7083	7096	7108	7120	7133	7145	7157	7169	7181	7193	44	12
46	7193	7206	7218	7230	7242	7254	7266	7278	7290	7302	7314	43	12
47	7314	7325	7337	7349	7361	7373	7385	7396	7408	7420	7431	42	12
48	7431	7443	7455	7466	7478	7490	7501	7513	7524	7536	7547	41	12
49	7547	7559	7570	7581	7593	7604	7615	7627	7638	7649	0.7660	40°	11
50°	0.7660	7672	7683	7694	7705	7716	7727	7738	7749	7760	7771	39	11
51	7771	7782	7793	7804	7815	7826	7837	7848	7859	7869	7880	38	11
52	7880	7891	7902	7912	7923	7934	7944	7955	7965	7976	7986	37	11
53	7986	7997	8007	8018	8028	8039	8049	8059	8070	8080	8090	36	10
54	8090	8100	8111	8121	8131	8141	8151	8161	8171	8181	0.8192	35	10
55	0.8192	8202	8211	8221	8231	8241	8251	8261	8271	8281	8290	34	10
56	8290	8300	8310	8320	8329	8339	8348	8358	8368	8377	8387	33	10
57	8387	8396	8406	8415	8425	8434	8443	8453	8462	8471	8480	32	9
58	8480	8490	8499	8508	8517	8526	8536	8545	8554	8563	8572	31	9
59	8572	8581	8590	8599	8607	8616	8625	8634	8643	8652	0.8660	30°	9
60°	0.8660	8669	8678	8686	8695	8704	8712	8721	8729	8738	8746	29	9
61	8746	8755	8763	8771	8780	8788	8796	8805	8813	8821	8829	28	8
62	8829	8838	8846	8854	8862	8870	8878	8886	8894	8902	8910	27	8
63	8910	8918	8926	8934	8942	8949	8957	8965	8973	8980	8988	26	8
64	8988	8996	9003	9011	9018	9026	9033	9041	9048	9056	0.9063	25	7
65	0.9063	9070	9078	9085	9092	9100	9107	9114	9121	9128	9135	24	7
66	9135	9143	9150	9157	9164	9171	9178	9184	9191	9198	9205	23	7
67	9205	9212	9219	9225	9232	9239	9245	9252	9259	9265	9272	22	7
68	9272	9278	9285	9291	9298	9304	9311	9317	9323	9330	9336	21	6
69	9336	9342	9348	9354	9361	9367	9373	9379	9385	9391	0.9397	20°	6
70°	0.9397	9403	9409	9415	9421	9426	9432	9438	9444	9449	9455	19	6
71	9455	9461	9466	9472	9478	9483	9489	9494	9500	9505	9511	18	6
72	9511	9516	9521	9527	9532	9537	9542	9548	9553	9558	9563	17	5
73	8563	9568	9573	9578	9583	9588	9593	9598	9603	9608	9613	16	5
74	9613	9617	9622	9627	9632	9636	9641	9646	9650	9655	0.9659	15	5
75	0.9659	9664	9668	9673	9677	9681	9686	9690	9694	9699	9703	14	4
76	9703	9707	9711	9715	9720	9724	9728	9732	9736	9740	9744	13	4
77	9744	9748	9751	9755	9759	9763	9767	9770	9774	9778	9781	12	4
78	9781	9785	9789	9792	9796	9799	9803	9806	9810	9813	9816	11	3
79	9816	9820	9823	9826	9829	9833	9836	9839	9842	9845	0.9848	10°	3
80°	0.9848	9851	9854	9857	9860	9863	9866	9869	9871	9874	9877	9	3
81	9877	9880	9882	9885	9888	9890	9893	9895	9898	9900	9903	8	3
82	9903	9905	9907	9910	9912	9914	9917	9919	9921	9923	9925	7	2
83	9925	9928	9930	9932	9934	9936	9938	9940	9942	9943	9945	6	2
84	9945	9947	9949	9951	9952	9954	9956	9957	9959	9960	0.9962	5	2
85	0.9962	9963	9965	9966	9968	9969	9971	9972	9973	9974	9976	4	1
86	9976	9977	9978	9979	9980	9981	9982	9983	9984	9985	9986	3	1
87	9986	9987	9988	9989	9990	9990	9991	9992	9993	9993	9994	2	1
88	9994	9995	9995	9996	9996	9997	9997	9997	9998	9998	0.9998	1	0
89	0.9998	9999	9999	9999	9999	0000	0000	0000	0000	0000	1.0000	0°	0
90°	1.0000												

	°.9 = (54')	°.8 (48')	°.7 (42')	°.6 (36')	°.5 (30')	°.4 (24')	°.3 (18')	°.2 (12')	°.1 (6')	°.0 (0')	Degrees

Natural Cosines

NATURAL TANGENTS AND COTANGENTS*

Natural Tangents at intervals of 0°.1, or 6'.

Degrees	°.0 =(0')	°.1 (6')	°.2 (12')	°.3 (18')	°.4 (24')	°.5 (30')	°.6 (36')	°.7 (42')	°.8 (48')	°.9 (54')			Average difference
											0.0000	90°	
0°	0.0000	0017	0035	0052	0070	0087	0105	0122	0140	0157	0175	89	17
1	0175	0192	0209	0227	0244	0262	0279	0297	0314	0332	0349	88	17
2	0349	0367	0384	0402	0419	0437	0454	0472	0489	0507	0524	87	17
3	0524	0542	0559	0577	0594	0612	0629	0647	0664	0682	0699	86	18
4	0699	0717	0734	0752	0769	0787	0805	0822	0840	0857	0.0875	85	18
5	0.0875	0892	0910	0928	0945	0963	0981	0998	1016	1033	1051	84	18
6	1051	1069	1086	1104	1122	1139	1157	1175	1192	1210	1228	83	18
7	1228	1246	1263	1281	1299	1317	1334	1352	1370	1388	1405	82	18
8	1405	1423	1441	1459	1477	1495	1512	1530	1548	1566	1584	81	18
9	1584	1602	1620	1638	1655	1673	1691	1709	1727	1745	0.1763	80°	18
10°	0.1763	1781	1799	1817	1835	1853	1871	1890	1908	1926	1944	79	18
11	1944	1962	1980	1998	2016	2035	2053	2071	2089	2107	2126	78	18
12	2126	2144	2162	2180	2199	2217	2235	2254	2272	2290	2309	77	18
13	2309	2327	2345	2364	2382	2401	2419	2438	2456	2475	2493	76	18
14	2493	2512	2530	2549	2568	2586	2605	2623	2642	2661	0.2679	75	19
15	0.2679	2698	2717	2736	2754	2773	2792	2811	2830	2849	2867	74	19
16	2867	2886	2905	2924	2943	2962	2981	3000	3019	3038	3057	73	19
17	3057	3076	3096	3115	3134	3153	3172	3191	3211	3230	3249	72	19
18	3249	3269	3288	3307	3327	3346	3365	3385	3404	3424	3443	71	19
19	3443	3463	3482	3502	3522	3541	3561	3581	3600	3620	0.3640	70°	20
20°	0.3640	3659	3679	3699	3719	3739	3759	3779	3799	3819	3839	69	20
21	3839	3859	3879	3899	3919	3939	3959	3979	4000	4020	4040	68	20
22	4040	4061	4081	4101	4122	4142	4163	4183	4204	4224	4245	67	21
23	4245	4265	4286	4307	4327	4348	4369	4390	4411	4431	4452	66	21
24	4452	4473	4494	4515	4536	4557	4578	4599	4621	4642	0.4663	65	21
25	0.4663	4684	4706	4727	4748	4770	4791	4813	4834	4856	4877	64	21
26	4877	4899	4921	4942	4964	4986	5008	5029	5051	5073	5095	63	22
27	5095	5117	5139	5161	5184	5206	5228	5250	5272	5295	5317	62	22
28	5317	5340	5362	5384	5407	5430	5452	5475	5498	5520	5543	61	23
29	5543	5566	5589	5612	5635	5658	5681	5704	5727	5750	0.5774	60°	23
30°	0.5774	5797	5820	5844	5867	5890	5914	5938	5961	5985	6009	59	24
31	6009	6032	6056	6080	6104	6128	6152	6176	6200	6224	6249	58	24
32	6249	6273	6297	6322	6346	6371	6395	6420	6445	6469	6494	57	25
33	6494	6519	6544	6569	6594	6619	6644	6669	6694	6720	6745	56	25
34	6745	6771	6796	6822	6847	6873	6899	6924	6950	6976	0.7002	55	26
35	0.7002	7028	7054	7080	7107	7133	7159	7186	7212	7239	7265	54	26
36	7265	7292	7319	7346	7373	7400	7427	7454	7481	7508	7536	53	27
37	7536	7563	7590	7618	7646	7673	7701	7729	7757	7785	7813	52	28
38	7813	7841	7869	7898	7926	7954	7983	8012	8040	8069	8098	51	28
39	8098	8127	8156	8185	8214	8243	8273	8302	8332	8361	0.8391	50°	29
40°	0.8391	8421	8451	8481	8511	8541	8571	8601	8632	8662	8693	49	30
41	8693	8724	8754	8785	8816	8847	8878	8910	8941	8972	9004	48	31
42	9004	9036	9067	9099	9131	9163	9195	9228	9260	9293	9325	47	32
43	9325	9358	9391	9424	9457	9490	9523	9556	9590	9623	0.9657	46	33
44	0.9657	9691	9725	9759	9793	9827	9861	9896	9930	9965	1.0000	45°	34
45°	1.0000												
	°.9 = (54')	°.8 (48')	°.7 (42')	°.6 (36')	°.5 (30')	°.4 (24')	°.3 (18')	°.2 (12')	°.1 (6')	°.0 (0')		Degrees	

Natural Cotangents

* From Marks, *Mechanical Engineers' Handbook*, 4th ed., McGraw-Hill, 1941.

NATURAL TANGENTS AND COTANGENTS (Continued)

Natural Tangents at intervals of 0°.1, or 6′.

Degrees	°.0 =(0′)	°.1 (6′)	°.2 (12′)	°.3 (18′)	°.4 (24′)	°.5 (30′)	°.6 (36′)	°.7 (42′)	°.8 (48′)	°.9 (54′)			Average difference
											1.0000	45°	
45°	1.0000	0035	0070	0105	0141	0176	0212	0247	0283	0319	0355	44	35
46	0355	0392	0428	0464	0501	0538	0575	0612	0649	0686	0724	43	37
47	0724	0761	0799	0837	0875	0913	0951	0990	1028	1067	1106	42	38
48	1106	1145	1184	1224	1263	1303	1343	1383	1423	1463	1504	41	40
49	1504	1544	1585	1626	1667	1708	1750	1792	1833	1875	1.1918	40°	41
50°	1.1918	1960	2002	2045	2088	2131	2174	2218	2261	2305	2349	39	43
51	2349	2393	2437	2482	2527	2572	2617	2662	2708	2753	2799	38	45
52	2799	2846	2892	2938	2985	3032	3079	3127	3175	3222	3270	37	47
53	3270	3319	3367	3416	3465	3514	3564	3613	3663	3713	3764	36	49
54	3764	3814	3865	3916	3968	4019	4071	4124	4176	4229	1.4281	35	52
55	1.4281	4335	4388	4442	4496	4550	4605	4659	4715	4770	4826	34	55
56	4826	4882	4938	4994	5051	5108	5166	5224	5282	5340	5399	33	57
57	5399	5458	5517	5577	5637	5697	5757	5818	5880	5941	6003	32	60
58	6003	6066	6128	6191	6255	6319	6383	6447	6512	6577	6643	31	64
59	1.6643	6709	6775	6842	6909	6977	7045	7113	7182	7251	1.7321	30°	67
60°	1.732	1.739	1.746	1.753	1.760	1.767	1.775	1.782	1.789	1.797	1.804	29	7
61	1.804	1.811	1.819	1.827	1.834	1.842	1.849	1.857	1.865	1.873	1.881	28	8
62	1.881	1.889	1.897	1.905	1.913	1.921	1.929	1.937	1.946	1.954	1.963	27	8
63	1.963	1.971	1.980	1.988	1.997	2.006	2.014	2.023	2.032	2.041	2.050	26	9
64	2.050	2.059	2.069	2.078	2.087	2.097	2.106	2.116	2.125	2.135	2.145	25	9
65	2.145	2.154	2.164	2.174	2.184	2.194	2.204	2.215	2.225	2.236	2.246	24	10
66	2.246	2.257	2.267	2.278	2.289	2.300	2.311	2.322	2.333	2.344	2.356	23	11
67	2.356	2.367	2.379	2.391	2.402	2.414	2.426	2.438	2.450	2.463	2.475	22	12
68	2.475	2.488	2.500	2.513	2.526	2.539	2.552	2.565	2.578	2.592	2.605	21	13
69	2.605	2.619	2.633	2.646	2.660	2.675	2.689	2.703	2.718	2.733	2.747	20°	14
70°	2.747	2.762	2.778	2.793	2.808	2.824	2.840	2.856	2.872	2.888	2.904	19	16
71	2.904	2.921	2.937	2.954	2.971	2.989	3.006	3.024	3.042	3.060	3.078	18	17
72	3.078	3.096	3.115	3.133	3.152	3.172	3.191	3.211	3.230	3.251	3.271	17	19
73	3.271	3.291	3.312	3.333	3.354	3.376	3.398	3.420	3.442	3.465	3.487	16	22
74	3.487	3.511	3.534	3.558	3.582	3.606	3.630	3.655	3.681	3.706	3.732	15	24
75	3.732	3.758	3.785	3.812	3.839	3.867	3.895	3.923	3.952	3.981	4.011	14	28
76	4.011	4.041	4.071	4.102	4.134	4.165	4.198	4.230	4.264	4.297	4.331	13	32
77	4.331	4.366	4.402	4.437	4.474	4.511	4.548	4.586	4.625	4.665	4.705	12	37
78	4.705	4.745	4.787	4.829	4.872	4.915	4.959	5.005	5.050	5.097	5.145	11	44
79	5.145	5.193	5.242	5.292	5.343	5.396	5.449	5.503	5.558	5.614	5.671	10°	53
80°	5.671	5.730	5.789	5.850	5.912	5.976	6.041	6.107	6.174	6.243	6.314	9	
81	6.314	6.386	6.460	6.535	6.612	6.691	6.772	6.855	6.940	7.026	7.115	8	
82	7.115	7.207	7.300	7.396	7.495	7.596	7.700	7.806	7.916	8.028	8.144	7	
83	8.144	8.264	8.386	8.513	8.643	8.777	8.915	9.058	9.205	9.357	9.514	6	
84	9.514	9.677	9.845	10.02	10.20	10.39	10.58	10.78	10.99	11.20	11.43	5	
85	11.43	11.66	11.91	12.16	12.43	12.71	13.00	13.30	13.62	13.95	14.30	4	
86	14.30	14.67	15.06	15.46	15.90	16.35	16.83	17.34	17.89	18.46	19.08	3	
87	19.08	19.74	20.45	21.20	22.02	22.90	23.86	24.90	26.03	27.27	28.64	2	
88	28.64	30.14	31.82	33.69	35.80	38.19	40.92	44.07	47.74	52.08	57.29	1	
89	57.29	63.66	71.62	81.85	95.49	114.6	143.2	191.0	286.5	573.0	∞	0°	
90°	∞												

| | °.9 = (54′) | °.8 (48′) | °.7 (42′) | °.6 (36′) | °.5 (30′) | °.4 (24′) | °.3 (18′) | °.2 (12′) | °.1 (6′) | °.0 (0′) | | Degrees | |

Natural Cotangents

DECIMAL EQUIVALENTS OF COMMON FRACTIONS*

8ths	16ths	32nds	64ths	Exact decimal values	8ths	16ths	32nds	64ths	Exact decimal values
			1	0.01 5625	4	8	16	32	0.50
		1	2	.03 125				33	.51 5625
			3	.04 6875				34	.53 125
	1	2	4	.06 25				35	.54 6875
			5	.07 8125		9	18	36	.56 25
		3	6	.09 375				37	.57 8125
			7	.10 9375			19	38	.59 375
1	2	4	8	.12 5				39	.60 9375
			9	.14 0625	5	10	20	40	.62 5
		5	10	.15 625				41	.64 0625
			11	.17 1875			21	42	.65 625
	3	6	12	.18 75				43	.67 1875
			13	.20 3125		11	22	44	.68 75
		7	14	.21 875				45	.70 3125
			15	.23 4375			23	46	.71 875
2	4	8	16	.25	6	12	24	47	.73 4375
			17	.26 5625	6	12	24	48	.75
		9	18	.28 125				49	.76 5625
			19	.29 6875			25	50	.78 125
	5	10	20	.31 25				51	.79 6875
			21	.32 8125		13	26	52	.81 25
		11	22	.34 375				53	.82 8125
			23	.35 9375			27	54	.84 375
3	6	12	24	.37 5	7	14	28	55	.85 9375
			25	.39 0625	7	14	28	56	.87 5
		13	26	.40 625				57	.89 0625
			27	.42 1875			29	58	.90 625
	7	14	28	.43 75				59	.92 1875
			29	.45 3125		15	30	60	.93 75
		15	30	.46 875				61	.95 3125
			31	.48 4375			31	62	.96 875
								63	.98 4375

STRENGTH OF U.S. STANDARD BOLTS FROM ¼ TO 3 IN. DIAMETER*

Bolt		Areas		Tensile strength, lb			Shearing strength, lb			
							Full bolt		Bottom of thread	
Diameter of bolt, in.	Number of threads per in.	Full bolt, sq in.	Bottom of thread, sq in.	At 10,000 lb per sq in.	At 12,500 lb per sq in.	At 17,500 lb per sq in.	At 7,500 lb per sq in.	At 10,000 lb per sq in.	At 7,500 lb per sq in.	At 10,000 lb per sq in.
¼	20	0.049	0.027	270	340	470	380	490	200	270
⁵⁄₁₆	18	0.077	0.045	450	570	790	580	770	340	450
³⁄₈	16	0.110	0.068	680	850	1,190	830	1,100	510	680
⁷⁄₁₆	14	0.150	0.093	930	1,170	1,630	1,130	1,500	700	930
½	13	0.196	0.126	1,260	1,570	2,200	1,470	1,960	940	1,260
⁹⁄₁₆	12	0.248	0.162	1,620	2,030	2,840	1,860	2,480	1,220	1,620
⁵⁄₈	11	0.307	0.202	2,020	2,520	3,530	2,300	3,070	1,510	2,020
¾	10	0.442	0.302	3,020	3,770	5,290	3,310	4,420	2,270	3,020
⅞	9	0.601	0.419	4,190	5,240	7,340	4,510	6,010	3,150	4,190
1	8	0.785	0.551	5,510	6,890	9,640	5,890	7,850	4,130	5,510
1⅛	7	0.994	0.693	6,990	8,660	12,130	7,450	9,940	5,200	6,930
1¼	7	1.227	0.890	8,890	11,120	15,570	9,200	12,270	6,670	8,900
1⅜	6	1,435	1.054	10,540	13,180	18,450	11,140	14,850	7,910	10,540
1½	6	1.767	1.294	12,940	16,170	22,640	13,250	17,670	9,700	12,940
1⅝	5½	2.074	1.515	15,150	18,940	26,510	15,550	20,740	11,360	15,150
1¾	5	2.405	1.745	17,450	21,800	30,520	18,040	24,050	13,080	17,440
1⅞	5	2.761	2.049	20,490	25,610	35,860	20,710	27,610	15,370	20,490
2	4½	3.142	2.300	23,000	28,750	40,500	23,560	31,420	17,250	23,000
2¼	4½	3.976	3.021	30,210	37,770	52,870	29,820	39,760	22,660	30,210
2½	4	4.909	3.716	37,160	46,450	65,040	36,820	49,090	27,870	37,160
2¾	4	5.940	4.620	46,200	57,750	80,840	44,580	59,400	34,650	46,200
3	3½	7.069	5.428	54,280	67,850	94,990	53,020	70,690	40,710	54,280

* From Marks, *Mechanical Engineers' Handbook*, 4th ed., McGraw-Hill, 1941.

SQUARE AND HEXAGONAL REGULAR BOLT HEADS*

(All dimensions in inches.)

Bolt diameter	Rough and semifinished					Finished				
	Width across flats		Min width across corners		Height	Width across flats		Min width across corners		Height
	Max	Min	Hex	Square		Max	Min	Hex	Square	
1/4	3/8	0.363	0.414	0.498	11/64	7/16	0.428	0.488	0.588	3/16
5/16	1/2	0.484	0.552	0.665	13/64	9/16	0.552	0.629	0.758	15/64
3/8	9/16	0.544	0.620	0.747	1/4	5/8	0.613	0.699	0.842	9/32
7/16	5/8	0.603	0.687	0.828	19/64	3/4	0.737	0.840	1.012	21/64
1/2	3/4	0.725	0.827	0.995	21/64	13/16	0.799	0.911	1.097	3/8
9/16	7/8	0.847	0.966	1.163	3/8	7/8	0.861	0.982	1.182	27/64
5/8	15/16	0.906	1.033	1.244	27/64	15/16	0.922	1.051	1.266	15/32
3/4	1 1/8	1.088	1.240	1.494	1/2	1 1/8	1.108	1.263	1.521	9/16
7/8	1 5/16	1.269	1.447	1.742	19/32	1 5/16	1.293	1.474	1.775	21/32
1	1 1/2	1.450	1.653	1.991	21/32	1 1/2	1.479	1.686	2.031	3/4
1 1/8	1 11/16	1.631	1.859	2.239	3/4	1 11/16	1.665	1.898	2.286	27/32
1 1/4	1 7/8	1.813	2.067	2.489	27/32	1 7/8	1.850	2.109	2.540	15/16
1 1/2	2 1/4	2.175	2.480	2.986	1	2 1/4	2.222	2.533	3.051	1 1/8
1 3/4	2 5/8	2.538	2.893	3.485	1 5/32	2 5/8	2.593	2.956	3.560	1 5/16
2	3	2.900	3.306	3.982	1 11/32	3	2.964	3.379	4.070	1 1/2
2 1/4	3 3/8	3.263	3.720	4.480	1 1/2	3 3/8	3.335	3.802	4.579	1 11/16
2 1/2	3 3/4	3.625	4.133	4.977	1 21/32	3 3/4	3.707	4.226	5.090	1 7/8
2 3/4	4 1/8	3.988	4.546	5.476	1 53/64	4 1/8	4.078	4.649	5.599	2 1/16
3	4 1/2	4.350	4.959	5.973	2	4 1/2	4.449	5.072	6.108	2 1/4

Regular nuts (rough, semifinished, and finished) have a maximum width across flats of $1\frac{1}{2}D$ except for $D = \frac{1}{4}$ to $\frac{9}{16}$ when the width = $1\frac{1}{2}D + \frac{1}{16}$. D is bolt diameter. Tolerance for width is $-0.050D$. Thickness is $\frac{7}{8}D$.

* From Marks, *Mechanical Engineers' Handbook*, 4th ed., McGraw-Hill, 1941.

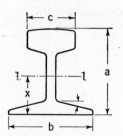

Standard rail section.

DIMENSIONS OF STANDARD RAILS AND LIGHT RAILS*

Standard and nominal weight	Weight per yd, lb	Area of section, sq in.	Dimensions, in.			Axis l–l			
			a	b	c	I, in.^{4}	r, in.	S, in.^{3}	x, in.
Penna. R.R. 152	152	.....	8	$6\frac34$	3	.....	....	.....	3.5
A.R.E.A. 131	130.8	12.86	$7\frac18$	6	3	88.5	2.65	22.6	3.2
A.R.E.A. 112	112.4	11.02	$6\frac58$	$5\frac12$	$2\frac{23}{32}$	65.8	2.44	18.1	2.98
A.S.C.E. 110	.....	.....	$6\frac18$	$6\frac18$	$2\frac78$				
A.R.E.A. 110	110.36	10.82	$6\frac14$	$5\frac12$	$2\frac{25}{32}$	57.0	2.30	16.8	$2\frac{53}{64}$
A.S.C.E. 100	100.3	9.84	$5\frac34$	$5\frac34$	$2\frac34$	44.0	2.11	14.6	2.73
A.A.R.–A 100	100.3	9.84	6	$5\frac12$	$2\frac34$	48.9	2.22	15.0	2.75
A.A.R.–B 100	.....	9.85	$5\frac{41}{64}$	$5\frac{9}{64}$	$2\frac{21}{32}$	41.3	2.05	13.7	2.63
A.R.E.A. 100	101.49	9.95	6	$5\frac38$	$2\frac34$	49.0	2.22	15.1	$2\frac34$
A.S.C.E. 95	94.5	9.28	$5\frac{9}{16}$	$5\frac{9}{16}$	$2\frac{11}{16}$	38.8	2.05	13.3	2.65
A.S.C.E. 90	89.9	8.83	$5\frac38$	$5\frac38$	$2\frac58$	34.4	1.97	12.2	2.55
A.A.R.–A 90	89.96	8.82	$5\frac18$	$5\frac18$	$2\frac{9}{16}$	38.7	2.19	12.5	$2\frac{35}{64}$
A.A.R.–B 90	90.7	8.87	$5\frac{17}{64}$	$5\frac{9}{64}$	$2\frac{21}{32}$	32.3	1.91	11.4	2.44
A.S.C.E. 85	84.7	8.33	$5\frac{3}{16}$	$5\frac{3}{16}$	$2\frac{9}{16}$	30.1	1.90	11.1	2.47
A.S.C.E. 80	80.0	7.86	5	5	$2\frac12$	26.4	1.83	10.1	2.38
A.A.R.–A 80	80.0	7.86	$5\frac18$	$4\frac58$	$2\frac12$	28.8	1.91	10.2	2.31
A.A.R.–B 80	80.8	7.91	$4\frac{15}{16}$	$4\frac{7}{16}$	$2\frac{7}{16}$	25.1	1.78	9.4	2.27
A.S.C.E. 75	74.6	7.33	$4\frac{13}{16}$	$4\frac{13}{16}$	$2\frac{15}{32}$	22.9	1.77	9.1	2.30
A.S.C.E. 70	69.5	6.81	$4\frac58$	$4\frac58$	$2\frac{7}{16}$	19.7	1.70	8.2	2.22
A.A.R.–A 70	69.7	6.82	$4\frac34$	$4\frac14$	$2\frac38$	21.0	1.75	8.2	2.20
A.A.R.–B 70	.....	.....	$4\frac{35}{64}$	$4\frac{36}{64}$	$2\frac38$				
A.S.C.E. 65	64.5	6.33	$4\frac{7}{16}$	$4\frac{7}{16}$	$2\frac{13}{32}$	16.9	1.63	7.4	2.14
A.S.C.E. 60	60.6	5.93	$4\frac14$	$4\frac14$	$2\frac38$	14.6	1.57	6.6	2.05
A.A.R.–A 60	60.0	5.86	$4\frac12$	4	$2\frac14$	15.4	1.62	6.5	2.13
A.A.R.–B 60	.....	.....	$4\frac{3}{16}$	$3\frac{11}{16}$	$2\frac18$				
A.S.C.E. 55	54.7	5.38	$4\frac{1}{16}$	$4\frac{1}{16}$	$2\frac14$	12.0	1.50	5.7	1.97
A.S.C.E. 50	49.6	4.87	$3\frac78$	$3\frac78$	$2\frac18$	9.9	1.43	5.0	1.88
A.S.C.E. 45	44.8	4.40	$3\frac{11}{16}$	$3\frac{11}{16}$	2	8.1	1.36	4.3	1.78
A.S.C.E. 40	40.1	3.94	$3\frac12$	$3\frac12$	$1\frac78$	6.6	1.29	3.6	1.68
A.S.C.E. 35	35.0	3.44	$3\frac{5}{16}$	$3\frac{5}{16}$	$1\frac34$	5.2	1.23	3.0	1.60
A.S.C.E. 30	30.5	3.00	$3\frac18$	$3\frac18$	$1\frac{11}{16}$	4.1	1.16	2.5	1.52
A.S.C.E. 25	24.3	2.39	$2\frac34$	$2\frac34$	$1\frac12$	2.5	1.02	1.8	1.33
A.S.C.E. 20	20.4	2.00	$2\frac58$	$2\frac58$	$1\frac{11}{32}$	1.9	0.99	1.4	1.27
A.S.C.E. 16	15.8	1.55	$2\frac38$	$2\frac38$	$1\frac{11}{64}$	1.2	0.89	1.0	1.15
A.S.C.E. 14	13.6	1.34	$2\frac{1}{16}$	$2\frac{1}{16}$	$1\frac{1}{16}$	0.76	0.75	0.73	1.02
A.S.C.E. 12	12.0	1.18	2	2	1	0.66	0.75	0.63	0.96
A.S.C.E. 10	9.76	0.96	$1\frac34$	$1\frac34$	$\frac{15}{16}$	0.40	0.65	0.46	0.87
A.S.C.E. 8	7.84	0.77	$1\frac{9}{16}$	$1\frac{9}{16}$	$\frac{13}{16}$	0.26	0.58	0.32	0.75

I = moment of inertia of section. r = radius of gyration. S = section modulus. x = distance from neutral axis to base.

* From Marks, *Mechanical Engineers' Handbook*, 4th ed., McGraw-Hill, 1941.

WIRE AND SHEET–METAL GAUGES*

(Diameters and thicknesses in decimal parts of an inch.)

Gauge No.	American wire gauge, or Brown & Sharpe (for nonferrous sheet and wire)	Steel wire gauge or Washburn & Moen or Roebling (for steel wire)	Birmingham wire gauge (B.W.G.) or Stubs iron wire (for steel rods or sheets)	Stubs steel wire gauge	British Imperial standard wire gauge (S.W.G.)	U.S. standard gauge for wrought iron sheet (480 lb per cu ft)	U.S. standard gauge for steel and open-hearth iron sheet (489.6 lb per cu ft)	British standard gauge for iron and steel sheets and hoops, 1914 (B.G.)
0000000	......	0.4900	......	......	0.500	0.500	0.4902	0.6666
000000	......	0.4615	......	......	0.464	0.469	0.4596	0.6250
00000	......	0.4305	......	......	0.432	0.438	0.4289	0.5883
0000	0.460	0.3938	0.454	......	0.400	0.406	0.3983	0.5416
000	0.410	0.3625	0.425	......	0.372	0.375	0.3676	0.5000
00	0.365	0.3310	0.380	......	0.348	0.344	0.3370	0.4452
0	0.325	0.3065	0.340	......	0.324	0.312	0.3064	0.3964
1	0.289	0.2830	0.300	0.227	0.300	0.281	0.2757	0.3532
2	0.258	0.2625	0.284	0.219	0.276	0.266	0.2604	0.3147
3	0.229	0.2437	0.259	0.212	0.252	0.250	0.2451	0.2804
4	0.204	0.2253	0.238	0.207	0.232	0.234	0.2298	0.2500
5	0.182	0.2070	0.220	0.204	0.212	0.219	0.2145	0.2225
6	0.162	0.1920	0.203	0.201	0.192	0.203	0.1991	0.1981
7	0.144	0.1770	0.180	0.199	0.176	0.188	0.1838	0.1764
8	0.128	0.1620	0.165	0.197	0.160	0.172	0.1685	0.1570
9	0.114	0.1483	0.148	0.194	0.144	0.156	0.1532	0.1398
10	0.102	0.1350	0.134	0.191	0.128	0.141	0.1379	0.1250
11	0.091	0.1205	0.120	0.188	0.116	0.125	0.1225	0.1113
12	0.081	0.1055	0.109	0.185	0.104	0.109	0.1072	0.0991
13	0.072	0.0915	0.095	0.182	0.092	0.094	0.0919	0.0882
14	0.064	0.0800	0.083	0.180	0.080	0.078	0.0766	0.0785
15	0.057	0.0720	0.072	0.178	0.072	0.070	0.0689	0.0699
16	0.051	0.0625	0.065	0.175	0.064	0.062	0.0613	0.0625
17	0.045	0.0540	0.058	0.172	0.056	0.056	0.0551	0.0556
18	0.040	0.0475	0.049	0.168	0.048	0.050	0.0490	0.0495
19	0.036	0.0410	0.042	0.164	0.040	0.0438	0.0429	0.0440
20	0.032	0.0348	0.035	0.161	0.036	0.0375	0.0368	0.0392
21	0.0285	0.0317	0.032	0.157	0.032	0.0344	0.0337	0.0349
22	0.0253	0.0286	0.028	0.155	0.028	0.0312	0.0306	0.0313
23	0.0226	0.0258	0.025	0.153	0.024	0.0281	0.0276	0.0278
24	0.0201	0.0230	0.022	0.151	0.022	0.0250	0.0245	0.0248
25	0.0179	0.0204	0.020	0.148	0.020	0.0219	0.0214	0.0220
26	0.0159	0.0181	0.018	0.146	0.018	0.0188	0.0184	0.0196
27	0.0142	0.0173	0.016	0.143	0.0164	0.0172	0.0169	0.0175
28	0.0126	0.0162	0.014	0.139	0.0148	0.0156	0.0153	0.0156
29	0.0113	0.0150	0.013	0.134	0.0136	0.0141	0.0138	0.0139
30	0.0100	0.0140	0.012	0.127	0.0124	0.0125	0.0123	0.0123
31	0.0089	0.0132	0.010	0.120	0.0116	0.0109	0.0107	0.0110
32	0.0080	0.0128	0.009	0.115	0.0108	0.0102	0.0100	0.0098
33	0.0071	0.0118	0.008	0.112	0.0100	0.0094	0.0092	0.0087
34	0.0063	0.0104	0.007	0.110	0.0092	0.0086	0.0084	0.0077
35	0.0056	0.0095	0.005	0.108	0.0084	0.0078	0.0077	0.0069
36	0.0050	0.0090	0.004	0.106	0.0076	0.0070	0.0069	0.0061
37	0.0045	0.0085	......	0.103	0.0068	0.0066	0.0065	0.0054
38	0.0040	0.0080	......	0.101	0.0060	0.0062	0.0061	0.0048
39	0.0035	0.0075	......	0.099	0.0052	0.0059	0.0057	0.0043
40	0.0031	0.0070	......	0.097	0.0048	0.0055	0.0054	0.0039
41	......	0.0066	......	0.095	0.0044	0.0053	0.0052	0.0034
42	......	0.0062	......	0.092	0.0040	0.0051	0.0050	0.0031
43	......	0.0060	......	0.088	0.0036	0.0049	0.0048	0.0027
44	......	0.0058	......	0.085	0.0032	0.0047	0.0046	0.0024
45	......	0.0055	......	0.081	0.0028	......	......	0.0022
46	......	0.0052	......	0.079	0.0024	......	......	0.0019
47	......	0.0050	......	0.077	0.0020	......	......	0.0017
48	......	0.0048	......	0.075	0.0016	......	......	0.0015
49	......	0.0046	......	0.072	0.0012	......	......	0.0014
50	......	0.0044	......	0.069	0.0010	......	......	0.0012

* From Marks, *Mechanical Engineers' Handbook*, 4th ed., McGraw-Hill, 1941.

WEIGHTS OF SQUARE AND ROUND STEEL BARS*

(For iron, subtract 2 per cent.)

Size, in.	Weight, lb per lin ft		Size, in.	Weight, lb per lin ft		Size, in.	Weight, lb per lin ft		Size, in.	Weight, lb per lin ft	
	Square	Round		Square	Round		Square	Round		Square	Round
0	.....	.....	3	30.60	24.03	6	122.4	96.1	9	275.4	216.3
1/16	0.013	0.010	1/16	31.89	25.05	1/16	125.0	98.2	1/16	279.2	219.3
1/8	0.053	0.042	1/8	33.20	26.08	1/8	127.6	100.2	1/8	283.1	222.4
3/16	0.120	0.094	3/16	34.54	27.13	3/16	130.2	102.2	3/16	287.0	225.4
1/4	0.213	0.167	1/4	35.91	28.21	1/4	132.8	104.3	1/4	290.9	228.5
5/16	0.332	0.261	5/16	37.31	29.30	5/16	135.5	106.4	5/16	294.9	231.6
3/8	0.478	0.376	3/8	38.73	30.42	3/8	138.2	108.5	3/8	298.8	234.7
7/16	0.651	0.511	7/16	40.18	31.55	7/16	140.9	110.7	7/16	302.8	237.8
1/2	0.850	0.668	1/2	41.65	32.71	1/2	143.7	112.8	1/2	306.9	241.0
9/16	1.076	0.845	9/16	43.15	33.89	9/16	146.4	115.0	9/16	310.9	244.2
5/8	1.328	1.043	5/8	44.68	35.09	5/8	149.2	117.2	5/8	315.0	247.4
11/16	1.607	1.262	11/16	46.23	36.31	11/16	152.1	119.4	11/16	319.1	250.6
3/4	1.913	1.502	3/4	47.81	37.55	3/4	154.9	121.7	3/4	323.2	253.9
13/16	2.245	1.763	13/16	49.42	38.81	13/16	157.8	123.9	13/16	327.4	257.1
7/8	2.603	2.044	7/8	51.05	40.10	7/8	160.7	126.2	7/8	331.6	260.4
15/16	2.988	2.347	15/16	52.71	41.40	15/16	163.6	128.5	15/16	335.8	263.7
1	3.400	2.670	4	54.40	42.73	7	166.6	130.9	10	340.0	267.0
1/16	3.838	3.015	1/16	56.11	44.07	1/16	169.6	133.2	1/16	344.3	270.4
1/8	4.303	3.380	1/8	57.85	45.44	1/8	172.6	135.6	1/8	348.6	273.8
3/16	4.795	3.766	3/16	59.62	46.83	3/16	175.6	137.9	3/16	352.9	277.1
1/4	5.313	4.172	1/4	61.41	48.23	1/4	178.7	140.4	1/4	357.2	280.6
5/16	5.857	4.600	5/16	63.23	49.66	5/16	181.8	142.8	5/16	361.6	284.0
3/8	6.428	5.049	3/8	65.08	51.11	3/8	184.9	145.2	3/8	366.0	287.4
7/16	7.026	5.518	7/16	66.95	52.58	7/16	188.1	147.7	7/16	370.4	290.9
1/2	7.650	6.008	1/2	68.85	54.07	1/2	191.3	150.2	1/2	374.9	294.4
9/16	8.301	6.519	9/16	70.78	55.59	9/16	194.5	152.7	9/16	379.3	297.9
5/8	8.978	7.051	5/8	72.73	57.12	5/8	197.7	155.3	5/8	383.8	301.5
11/16	9.682	7.604	11/16	74.71	58.67	11/16	200.9	157.8	11/16	388.4	305.0
3/4	10.413	8.178	3/4	76.71	60.25	3/4	204.2	160.4	3/4	392.9	308.6
13/16	11.170	8.773	13/16	78.74	61.85	13/16	207.5	163.0	13/16	397.5	312.2
7/8	11.953	9.388	7/8	80.80	63.46	7/8	210.9	165.6	7/8	402.1	315.8
15/16	12.763	10.024	15/16	82.89	65.10	15/16	214.2	168.2	15/16	406.7	319.5
2	13.600	10.681	5	85.00	66.76	8	217.6	170.9	11	411.4	323.1
1/16	14.463	11.359	1/16	87.14	68.44	1/16	221.0	173.6	1/16	416.1	326.8
1/8	15.353	12.058	1/8	89.30	70.14	1/8	224.5	176.3	1/8	420.8	330.5
3/16	16.270	12.778	3/16	91.49	71.86	3/16	227.9	179.0	3/16	425.5	334.2
1/4	17.213	13.519	1/4	93.71	73.60	1/4	231.4	181.8	1/4	430.3	338.0
5/16	18.182	14.280	5/16	95.96	75.36	5/16	234.9	184.5	5/16	435.1	341.7
3/8	19.178	15.062	3/8	98.23	77.15	3/8	238.5	187.3	3/8	439.9	345.5
7/16	20.201	15.866	7/16	100.53	78.95	7/16	242.1	190.1	7/16	444.8	349.3
1/2	21.250	16.690	1/2	102.85	80.78	1/2	245.7	192.9	1/2	449.7	353.2
9/16	22.326	17.534	9/16	105.20	82.62	9/16	249.3	195.8	9/16	454.6	357.0
5/8	23.428	18.400	5/8	107.58	84.49	5/8	252.9	198.7	5/8	459.5	360.9
11/16	24.557	19.287	11/16	109.98	86.38	11/16	256.6	201.5	11/16	464.4	364.8
3/4	25.713	20.195	3/4	112.41	88.29	3/4	260.3	204.5	3/4	469.4	368.7
13/16	26.895	21.123	13/16	114.87	90.22	13/16	264.0	207.4	13/16	474.4	372.6
7/8	28.103	22.072	7/8	117.35	92.17	7/8	267.8	210.3	7/8	479.5	376.6
15/16	29.338	23.042	15/16	119.86	94.14	15/16	271.6	213.3	15/16	484.5	380.5

* From Marks, *Mechanical Engineers' Handbook*, 4th ed., McGraw-Hill, 1941.

STANDARD PIPE AND LINE PIPE*

Nominal internal, in.	Diameter — Actual external, in.	Diameter — Approx internal diam, in.	Nominal thickness, in.	Circumference — External, in.	Circumference — Internal, in.	Transverse areas — External, sq in.	Transverse areas — Internal, sq in.	Transverse areas — Metal, sq in.	Length of pipe per sq ft of — External surface, ft	Length of pipe per sq ft of — Internal surface, ft	Length of pipe containing 1 cu ft, ft	Nominal weight per ft, lb	No. of threads per in. of screw
⅛	0.405	0.27	0.068	1.27	0.85	0.13	0.06	0.07	9.44	14.15	2513.00	0.24	27
¼	0.540	0.36	0.088	1.70	1.14	0.23	0.10	0.12	7.08	10.49	1383.30	0.42	18
⅜	0.675	0.49	0.091	2.12	1.55	0.36	0.19	0.17	5.66	7.76	751.20	0.57	18
½	0.840	0.62	0.109	2.63	1.95	0.55	0.30	0.25	4.55	6.15	472.40	0.85	14
¾	1.050	0.82	0.113	3.30	2.59	0.87	0.53	0.33	3.64	4.64	270.00	1.13	14
1	1.315	1.05	0.134	4.13	3.29	1.36	0.86	0.50	2.90	3.65	166.90	1.68	11½
1¼	1.660	1.38	0.140	5.22	4.34	2.16	1.50	0.67	2.30	2.77	96.25	2.27	11½
1½	1.900	1.61	0.145	5.97	5.06	2.84	2.04	0.80	2.01	2.37	70.66	2.72	11½
2	2.375	2.07	0.154	7.46	6.49	4.43	3.36	1.07	1.61	1.85	42.91	3.65	11½
2½	2.875	2.47	0.204	9.03	7.75	6.49	4.78	1.71	1.33	1.55	30.10	5.79	8
3	3.500	3.07	0.217	11.00	9.63	9.62	7.39	2.24	1.09	1.25	19.50	7.57	8
3½	4.000	3.55	0.226	12.57	11.15	12.57	9.89	2.68	0.96	1.08	14.57	9.11	8
4	4.500	4.03	0.237	14.14	12.65	15.90	12.73	3.18	0.85	0.95	11.31	10.79	8
5	5.563	5.05	0.259	17.48	15.85	24.31	19.99	4.32	0.69	0.76	7.20	14.62	8
6	6.625	6.07	0.280	20.81	19.05	34.47	28.89	5.59	0.58	0.63	4.98	18.97	8
8	8.625	8.07	0.276	27.10	25.35	58.43	51.15	7.28	0.44	0.47	2.82	24.69	8
8	8.625	7.98	0.322	27.10	25.07	58.43	50.02	8.41	0.44	0.48	2.88	28.55	8
9	9.625	8.94	0.344	30.24	28.08	72.76	62.72	10.04	0.40	0.43	2.29	33.91	8
10	10.750	10.19	0.278	33.77	32.01	90.76	81.55	9.21	0.36	0.37	1.76	31.20	8
10	10.750	10.14	0.306	33.77	31.86	90.76	80.75	10.01	0.36	0.38	1.78	34.24	8
10	10.750	10.02	0.366	33.77	31.47	90.76	78.82	11.94	0.36	0.38	1.82	40.48	8
12	12.750	12.09	0.328	40.06	37.98	127.68	114.80	12.88	0.30	0.32	1.25	43.77	8
12	12.750	12.00	0.375	40.06	37.70	127.68	113.10	14.59	0.30	0.32	1.27	49.56	8

EXTRA-STRONG PIPE*

Nominal internal, in.	Diameter — Actual external, in.	Diameter — Approx internal diam, in.	Nominal thickness, in.	Circumference — External, in.	Circumference — Internal, in.	Transverse areas — External, sq in.	Transverse areas — Internal, sq in.	Transverse areas — Metal, sq in.	Length of pipe per sq ft of — External surface, ft	Length of pipe per sq ft of — Internal surface, ft	Nominal weight per ft, lb
⅛	0.405	0.21	0.100	1.27	0.64	0.13	0.03	0.10	9.43	18.63	0.31
¼	0.540	0.29	0.123	1.70	0.92	0.23	0.07	0.16	7.08	12.99	0.54
⅜	0.675	0.42	0.127	2.12	1.32	0.36	0.14	0.22	5.66	9.07	0.74
½	0.840	0.54	0.149	2.64	1.70	0.55	0.23	0.32	4.55	7.05	1.09
¾	1.050	0.74	0.157	3.30	2.31	0.87	0.43	0.44	3.64	5.11	1.47
1	1.315	0.95	0.182	4.13	2.99	1.36	0.71	0.65	2.90	4.02	2.17
1¼	1.660	1.27	0.194	5.22	4.00	2.16	1.27	0.89	2.30	3.00	2.99
1½	1.900	1.49	0.203	5.97	4.69	2.84	1.75	1.08	2.01	2.56	3.63
2	2.375	1.93	0.221	7.46	6.07	4.43	2.94	1.50	1.61	1.98	5.02
2½	2.875	2.32	0.280	9.03	7.27	6.49	4.21	2.28	1.33	1.65	7.66
3	3.500	2.89	0.304	11.00	9.09	9.62	6.57	3.05	1.09	1.33	10.25
3½	4.000	3.36	0.321	12.57	10.55	12.57	8.86	3.71	0.96	1.14	12.50
4	4.500	3.82	0.341	14.14	12.00	15.90	11.45	4.46	0.85	1.00	14.98
5	5.563	4.81	0.375	17.48	15.12	24.31	18.19	6.11	0.69	0.79	20.78
6	6.625	5.75	0.437	20.81	18.07	34.47	25.98	8.50	0.58	0.66	28.57
8	8.625	7.63	0.500	27.10	23.96	58.43	45.66	12.76	0.44	0.50	43.34
10	10.750	9.75	0.500	33.77	30.63	90.76	74.66	16.10	0.36	0.40	54.73
12	12.750	11.75	0.500	40.06	36.91	127.68	108.43	19.25	0.30	0.33	65.41

* From Marks, *Mechanical Engineers' Handbook*, 4th ed., McGraw-Hill, 1941.

DOUBLE-EXTRA-STRONG PIPE*

Nominal pipe size, in.	Outside diameter, in.	Nominal wall thickness, in.		Wt per ft, lb, plain ends	Nominal pipe size, in.	Outside diameter, in.	Nominal wall thickness, in.		Wt per ft, lb, plain ends
		Wrought iron	Steel				Wrought iron	Steel	
½	0.840	0.307	0.294	1.714	2½	2.875	0.565	0.552	13.695
¾	1.050	0.318	0.308	2.440	3	3.500	0.615	0.600	18.583
1	1.315	0.369	0.358	3.659	4	4.500	0.690	0.674	27.541
1¼	1.660	0.393	0.382	5.214	5	5.563	0.768	0.750	38.552
1½	1.900	0.411	0.400	6.408	6	6.625	0.884	0.864	53.160
2	2.375	0.447	0.436	9.029	8	8.625	0.895	0.875	72.424

* From Marks, *Mechanical Engineers' Handbook*, 4th ed., McGraw-Hill, 1941.

ALLOWABLE STRUCTURAL STRESSES
New York City Building Code,
Volume I, Section C26-368.0

Tension

Structural members.................. 20,000 psi

Unfinished bolts (root diam)........... 15,000

High strength bolts (root diam)........ 20,000

Rivets............................ 20,000

Butt welds......................... 20,000

Bending

Structural members.................. 20,000 psi

Unbraced structural members.......... $\dfrac{12,000,000bt}{Ld}$

where $L = \text{more than } \dfrac{600bt}{d}$

Compression

Very short members.................. 20,000 psi

Members where L/r less than 120...... $17,000 - 0.485 \times \dfrac{L^2}{r^2}$

Members where L/r is 120 to 200...... $\dfrac{18,000}{1 + \dfrac{L^2}{18,000r^2}}$

Shear

Unfinished bolts.................... 10,000 psi

Pins, rivets, high strength bolts........ 15,000

Beam web, gross area................ 13,000

Fillet weld (at throat)............... 13,600

Bearing

Pins.....	single shear—32,000 psi	double shear—32,000 psi
Rivets...	single shear—40,000	double shear—32,000
Bolts.....	single shear—25,000	double shear—20,000

Combined stresses

Axial plus bending

$$\frac{\text{actual axial stress}}{\text{allowable compressive stress}} + \frac{\text{actual bending stress}}{\text{allowable bending stress}}$$

must be less than 1.

INDEX